Manual of Water Supply Practices

M23

PVC Pipe—Design and Installation

Third Edition

Erratum incorporated October 2022.

American Water Works Association

Manual of Water Supply Practices—M23, Third Edition

PVC Pipe—Design and Installation

If you find errors in this manual, please email books@awwa.org. Possible errata will be posted at www.awwa.org/resources-tools/resource.development.groups/manuals-program.aspx.

Senior Managing Editor/Project Manager: Melissa Valentine
Technical Editor: Suzanne Snyder
Cover Design/Technical Illustrations: Michael Labruyere
Manager, Publishing Operations: Gillian Wink
Compositor: Innodata
Senior Specialist - Manuals: Willadee Hitchcock
Manuals Coordinator 2: Janet Greifinger

Library of Congress Cataloging-in-Publication Data

Names: Walker, Robert (Hydraulic engineer), author. | American Water Works Association, editor.
Title: M23 PVC pipe : design and installation / by Robert Walker.
Other titles: PVC pipe | PVC pipe--design and installation.
Description: Third edition. | Denver, CO : American Water Works Association, [2020] | Revised edition of: PVC pipe--design and installation. | Includes bibliographical references and index. | Summary: "This manual provides the user with both general and technical information to aid in design, procurement, installation, and maintenance of PVC pipe and fittings. This manual presents a discussion of recommended practices"-- Provided by publisher.
Identifiers: LCCN 2020000645 | ISBN 9781625763600 (paperback)
Subjects: LCSH: Water-pipes--Design and construction. | Pipe, Plastic. | Polyvinyl chloride.
Classification: LCC TD491 .W28 2020 | DDC 696/.1--dc23
LC record available at https://lccn.loc.gov/2020000645

Printed in the United States of America

ISBN 978-1-62576-360-0 Electronic ISBN-13 978-1-61300-530-9

DOI https://doi.org/10.12999/AWWA.M23ed3

American Water Works Association

American Water Works Association
6666 West Quincy Avenue
Denver, CO 80235-3098
awwa.org

Contents

Figures

Tables

Preface

This is the third edition of AWWA M23, *PVC Pipe—Design and Installation*. This manual provides the user with both general and technical information to aid in design, procurement, installation, and maintenance of PVC pipe and fittings.

This manual presents a discussion of recommended practices. It is not intended to be a technical commentary on AWWA standards that apply to PVC pipe, fittings, and related appurtenances.

This page intentionally blank.

Acknowledgments

The following key individuals and members of the PVC Pipe Design Manual Subcommittee were instrumental in updating this edition of the M23 manual.

Robert Walker, (Task Group Chairman) Underground Solutions, Southlake, Tex.
Linda Bowles, Bureau of Reclamation, Denver, Colo.
Amster Howard, Consultant, Lakewood, Colo.
Tim McCandless, Staff Engineer Liaison, AWWA, Denver, Colo.
Jay Parvez, Uni-Bell PVC Pipe Association, Dallas, Tex.
Larry Schmidt, Diamond Plastics Corporation, Grand Island, Neb.
Suzanne Scott, North American Pipe Corporation, Woodbridge, Ont. Canada
Dennis Shumard, EBAA Iron, Katy, Tex.
Bill Whidden, Woolpert, Orlando, Fla.

This manual was developed by the AWWA Standards Committee on PVC Pressure Pipe and Fittings. The membership of the committee at the time it approved this manual was as follows:

W. R. Whidden, (Committee Chair) Woolpert, Orlando, Fla.
S. D. Adams, Gwinnett County Dept. of Water Resources, Lawrenceville, Ga.
J. R. Barbier, Denver Water, Denver, Colo.
R. R. Bishop, Diamond Plastics Corporation, Grand Island, Neb
L. M. Bowles, (Chapter Chair) Bureau of Reclamation, Denver, Colo.
J. P. Castronovo, Blue Ridge, Ga.
A. Chastain-Howley, (Standards Council Liaison) Atonix Digital, Frisco, Tex.
A. J. Ciechanowski, NSF International, Ann Arbor, Mich.
S. J. Cook, Black & Veatch, Virginia Beach, Va.
S. E. Cooper, Uni-Bell PVC Pipe Association, Louisville, Ky.
S. Ferry, PSILab Inc., Longmont, Colo.
G. Garcia, JM Eagle, Los Angeles, Calif.
L. J. Gill, (Chapter Chair) Ipex Management Inc., Oakville, Ont. Canada
S.B. Gross, North American Pipe, Wayne, Pa.
G. Gundel, Specified Fittings, Bellingham, Wash.
J. Hebenstreit, UL LLC, Northbrook, Ill.
W. Hitchcock, (Staff Adviser) AWWA, Denver, Colo.
J. F. Houle, John F. Houle LLC, Livingston, Tex.
M. Huynh, JM Manufacturing Company Inc. DBA JM Eagle, Los Angeles, Calif.
K. S. Jeng-Bulloch, City of Houston, Houston, Tex.
A. Korell, City of North Bay, North Bay, Ont. Canada
J. Larson, Marshall Municipal Utilities, Marshall, Minn.
T. E. Layton, Orange County Utilities, Orlando, Fla.
S. C. Macleod, UL LLC, Melville, N.Y.
M. T. Marino, Nussbaumer & Clarke, Inc., Buffalo, N.Y.

A. Marner, Bureau of Reclamation, Denver, Colo.
T. Marti, (Chapter Chair) Underground Solutions Inc., Warrendale, Pa.
T.J. McCandless, (Staff Engineer) AWWA, Denver, Colo.
S. A. McKelvie, HDR Engineering Inc., Boston, Mass.
J. R. Paschal, Paschal Engineering, Sandy, Utah
M. Rennau, Diamond Plastics Corporation, Grand Island, Neb.
J. Riordan, HARCO Fittings, Lynchburg, Va.
J. Rudolf, Diamond Plastics Corporation, Grand Island, Neb.
V. B. Sandoval, Natural Resources Cons. Svc., Boise, Idaho
L. D. Schmidt, (Chapter Chair) Diamond Plastics Corporation, Grand Island, Neb.
S. Scott, (Chapter Chair) Royal Building Products, Woodbridge, Ont. Canada
D. Shumard, EBAA Iron, Katy, Tex.
J. K. Snyder, Snyder Environmental Engineering Associates, Audubon, Pa.
B. Sukolsky, Specified Fittings, Bellingham, Wash.
S. Titus, Washington Suburban Sanitary Commission, Laurel, Md.
R. P. Walker, (Chapter Chair) Underground Solutions, Southlake, Tex.

Chapter 1

General Properties of Polyvinyl Chloride Pipe

Polyvinyl chloride (PVC)* was discovered in the late nineteenth century. Scientists at that time found the new plastic material unusual in that it appeared nearly inert to most chemicals. However, it was soon discovered that the material was resistant to change, and it was concluded that the material could not be easily formed or processed into usable applications.

In the 1920s, scientific curiosity again brought PVC to public attention. In Europe and America, extended efforts eventually brought PVC plastics to the modern world. Technology, worldwide and particularly in Germany, slowly evolved for the use of PVC in its unplasticized, rigid form, which today is used in the production of a great many extruded and molded products. In the mid-1930s, German scientists and engineers developed and produced limited quantities of PVC pipe. Some PVC pipe installed at that time continues to provide satisfactory service today. Molecularly oriented polyvinyl chloride (PVCO) pressure pipe has been installed in Europe since the early 1970s and in North America since 1991. Fusible PVC pipe has been installed in North America since 2004.

MATERIAL PROPERTIES OF PVC PIPE COMPOUNDS

Polyvinyl chloride pipe and fabricated fittings derive properties and characteristics from the properties of their raw material components. Essentially, PVC pipe and fabricated

* Unless otherwise noted in this manual, "polyvinyl chloride pipe" or "PVC" refers to PVC bell-and-spigot pipe and PVC fusible pipe; "PVCO" refers to molecularly oriented PVC pipe.

fittings are manufactured from PVC extrusion compounds. Injection molded fittings use slightly different molding compounds. PVCO is manufactured from conventional PVC extrusion compounds. The following summary of the material properties for these compounds provides a solid foundation for an understanding and appreciation of PVC pipe properties.

Polyvinyl chloride resin, the basic building block of PVC pipe, is a polymer derived from air, salt water, and natural gas. PVC resin, produced by any of the common manufacturing processes (bulk, suspension, or emulsion), is combined with heat stabilizers, lubricants, and other ingredients to make PVC compound that can be extruded into pipe or molded into fittings.

Chemical and taste-and-odor evaluations of PVC compounds for potable water conveyance are conducted in accordance with procedures established by NSF International.* The extracted water must not exceed the maximum contaminant levels established by the US Environmental Protection Agency's (USEPA's) National Interim Primary Drinking Water Regulations (1975) and by the NSF limits of acceptance for residual vinyl chloride monomer and for taste and odor as shown in Table 1-1 of NSF Standard 61. Monitoring is conducted by NSF International or approved laboratories.

PVC pipe extrusion compounds must provide acceptable design stress properties as determined by long-term testing under hydrostatic pressure. Hydrostatic design stress ratings for pipe compounds are established after 10,000 hr of hydrostatic testing. Long-term performance of injection molded PVC fittings compounds are subject to at least 2,000 hr of hydrostatic testing.

AWWA's† PVC pipe and fittings standards define the basic properties of PVC compound using the American Society for Testing and Materials‡ (ASTM) Specification D1784, Standard Specification for Rigid Poly(Vinyl Chloride) (PVC) Compounds and Chlorinated Poly(Vinyl Chloride) (CPVC) Compounds. The specification includes a five-digit cell class designation system by which PVC compounds are classified according to their physical properties.

As shown in Table 1-1, the five properties designated are (1) base resin, (2) impact strength, (3) tensile strength, (4) elastic modulus in tension, and (5) deflection temperature under loading. Figure 1-1 shows how the classification system establishes minimum properties for the compound 12454, which is used in PVC pressure pipe manufactured in accordance with ANSI/AWWA C900, Polyvinyl Chloride (PVC) Pressure Pipe and Fabricated Fittings, 4 In. Through 60 In. (100 mm Through 1500 mm) and ANSI/AWWA C909, Molecularly Oriented Polyvinyl Chloride (PVCO) Pressure Pipe, 4 In. Through 24 In. (100 mm Through 600 mm) for Water, Wastewater, and Reclaimed Water Service. The material classification can be found on the pipe as part of its identification marking. Cell class is not part of the required marking in C900 and C909.

Many of the important properties of PVC pipe are predetermined by the characteristics of the PVC compound from which the pipe is extruded. PVC pressure pipe manufactured in accordance with ANSI/AWWA C900 or C909 must be extruded from PVC compound with cell classification 12454 or better. Those compounds must also qualify for a hydrostatic design basis of 4,000 psi (27.58 MPa) for water at 73.4°F (23°C) per the requirements of PPI§ TR-3, Policies and Procedures for Developing Hydrostatic Design Basis (HDB), Pressure Design Basis (PDB), Strength Design Basis (SDB), and Minimum Required Strength (MRS) Ratings for Thermoplastic Piping Materials or Pipe.

---

* NSF International, P.O. Box 130140, 789 N. Dixboro Rd., Ann Arbor, MI 48105.

† AWWA, 6666 W. Quincy Ave., Denver, CO 80235.

‡ ASTM, 100 Barr Harbor Drive, P.O. Box C700, West Conshohocken, PA 19428.

§ Plastics Pipe Institute, 105 Decker Court, Suite 825, Irving, TX 75062.

Table 1-1 Cell class requirements for rigid poly(vinyl chloride) compounds*

Order No.	Designation Property and Unit	Cell Limits 0	1	2	3	4	5	6	7	8
1	Base resin	Unspecified	Poly(vinyl chloride) homopolymer	Chlorinated poly(vinyl chloride)	Ethylene vinyl chloride copolymer	Propylene vinyl chloride copolymer	Vinyl acetate-vinyl chloride copolymer	Alkyl vinyl ether-vinyl chloride copolymer		
2	Impact strength (Izod), min.									
	J/m of notch	Unspecified	<34.7	34.7	80.1	266.9	533.8	800.7		
	ft-lb/in. of notch		<0.65	0.65	1.5	5.0	10.0	15.0		
3	Tensile strength, min.									
	MPa	Unspecified	<34.5	34.5	41.4	48.3	55.2			
	psi		<5,000	5,000	6,000	7,000	8,000			
4	Modulus of elasticity in tension, min.									
	MPa	Unspecified	<1,930	1,930	2,206	2,482	2,758	3,034		
	psi		<280,000	280,000	320,000	360,000	400,000	440,000		
5	Deflection temperature under load, min. 1.82 MPa (264 psi):									
	deg C	Unspecified	<55	55	60	70	80	90	100	110
	deg F		<131	131	140	158	176	194	212	230

Source: ASTM D1784, ASTM International, 100 Barr Harbor Drive, P.O. Box C700, West Conshohocken, PA 19428-2959.

* The minimum (min.) property value will determine the cell number, although the maximum expected value may fall within a higher cell.

Note: Flammability. All compounds covered by this specification, when tested in accordance with method D635, shall yield the following results: average extent of burning of <25 mm; average time of burning of <10 s.

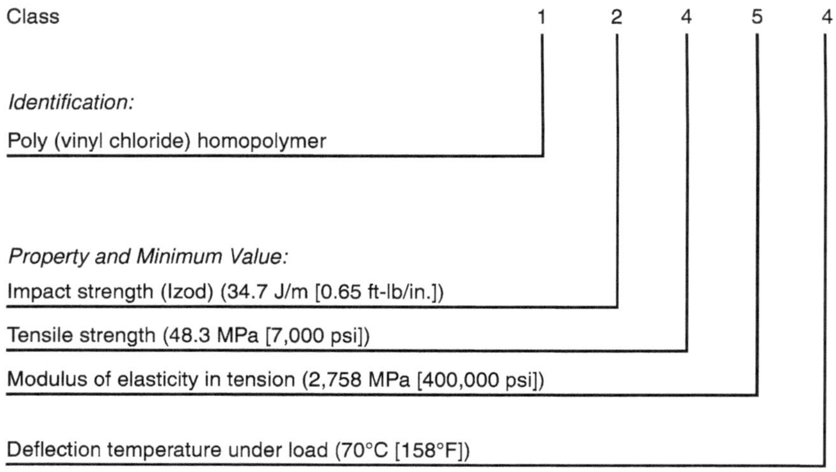

Source: ASTM D1784, ASTM International, 100 Barr Harbor Drive, P.O. Box C700, West Conshohocken, PA 19428-2959.
Note: The cell-type format provides the means for identification and close characterization and specification of material properties, alone or in combination, for a broad range of materials. This type format, however, is subject to possible misapplication since unobtainable property combinations can be selected if the user is not familiar with commercially available materials. The manufacturer should be consulted.

Figure 1-1 Class 12454 requirements

The manner in which selected materials are identified by this classification system is illustrated by a Class 12454 rigid PVC compound having the requirements shown in Table 1-1 and Figure 1-1.

CORROSION, PERMEATION, AND CHEMICAL RESISTANCE

PVC and PVCO pipes are resistant to almost all types of corrosion—both chemical and electrochemical—that are experienced in underground piping systems. Because PVC is a nonconductor, galvanic and electrochemical effects are nonexistent in PVC piping systems. PVC pipe cannot be damaged by aggressive waters or corrosive soils. Consequently, no lining, coating, cathodic protection, or plastic encasement is required when PVC and PVCO pipes are used.

Permeation

The selection of materials is critical for water service and distribution piping in locations where the pipe may be exposed to significant concentrations of pollutants comprised of low molecular weight petroleum products or organic solvents or their vapors. If a water pipe must pass through an area subject to contamination, the manufacturer should be consulted regarding permeation of pipe walls, jointing materials, etc., *before* selecting materials for use in that area.

When the contaminant of concern is commercial gasoline, the Water Research Foundation project, "Impact of Hydrocarbons on PE/PVC Pipes and Pipe Gaskets," offers considerable design guidance. One of the key findings of the report is that the PVC pipe wall is impervious to gasoline, either the free product, the vapor, or when it has contaminated the groundwater (p. 174 of the report). The most commonly used gasket (SBR) for the gasket PVC pipe joint was also studied. The report states that SBR gaskets can be used safely in water mains in contact with groundwater contaminated with any level of gasoline (p. 173). The report also states that SBR gaskets may be used (without any contaminant exceeding its maximum contaminant level) for water mains in all conditions of gasoline contamination provided there is at least a minimal flow in the main (e.g., the velocity is 2 ft/s or greater).

Chemical Resistance

Pipe. Response of PVC pipe under normal conditions to commonly anticipated chemical exposures is shown in Table 3.1 of the *Uni-Bell Handbook of PVC Pipe —Design and Construction.* Resistance of PVC pipe to reaction with or attack by the chemical substances listed has been determined by research and investigation. The information is primarily based on the immersion of unstressed strips into the chemicals and, to a lesser degree, on field experience. In most cases, the detailed test conditions, such as stress, exposure time, change in weight, change in volume, and change in strength, were not reported. Because of the complexity of some organochemical reactions, additional long-term testing should be performed for critical applications. Data provided are intended only as a guide and should not necessarily be regarded as applicable to all exposure durations, concentrations, or working conditions. The chemical resistance data are similar for PVCO pipe.

Gaskets. A check of the chemical resistance of the gasket should be completed independently of that for the pipe. Because gasket and pipe materials are different, so too are their abilities to resist chemical attack. Similarly, charts for resistance of gasket materials to chemical attack are based on manufacturers' testing and experience. The use of these charts is complicated by the fact that more than one elastomer may be present in a rubber compound. Chemical resistance information for commonly used gasket materials is provided in Table 3-2 and 3.3 of the *Uni-Bell Handbook of PVC Pipe—Design and Construction.*

Tables 3.2 and 3.3 provide general guidelines regarding the suitability of various elastomers currently used in these chemicals and services. The ratings are primarily based on literature published by various polymer suppliers and rubber manufacturers, as well as the opinions of experienced compounders. Several factors must be considered in using a rubber or polymer part. The most important of these factors include the following:

- Temperature of service. Higher temperatures increase the effect of all chemicals on polymers. The increase varies with the polymer and the chemical. A compound quite suitable at room temperature may perform poorly at elevated temperatures.
- Conditions of service. A compound that swells considerably might still function well as a static seal yet fail in any dynamic application.
- Grade of the polymer. Many types of polymers are available in different grades that vary greatly in chemical resistance.
- The compound itself. Compounds designed for other outstanding properties may be poorer in performance in a chemical than one designed especially for fluid resistance.
- Availability. Consult the elastomer manufacturers for availability of a compound for use as a PVC pipe gasket material.

If it is anticipated that gasket elastomers will be exposed to aggressive chemicals, it is advisable to test the elastomers.

ENVIRONMENTAL EFFECTS

The following paragraphs discuss the effects of environmental factors on PVC pipe, including temperature, biological attack, weather, abrasion, and tuberculation.

Thermal Effects

The performance of PVC pipe is significantly related to its operating temperature. Because it is a thermoplastic material, PVC will display variations in its physical properties as temperature changes (Figure 1-2). PVC pipe can be installed properly over the ambient temperature range in which construction crews can work. PVC pipe is rated for performance properties at a temperature of 73.4°F (23°C); however, it is recognized that operating temperatures of 33–90°F (1–32°C) do exist in water systems. As the operating temperature decreases, the pipe's stiffness and tensile strength increase, thereby increasing the pipe's pressure capacity and its ability to resist earth-loading deflection. At the same time, PVC pipe loses impact strength and becomes less ductile as temperature decreases, necessitating greater handling care in sub-freezing weather. As the operating temperature

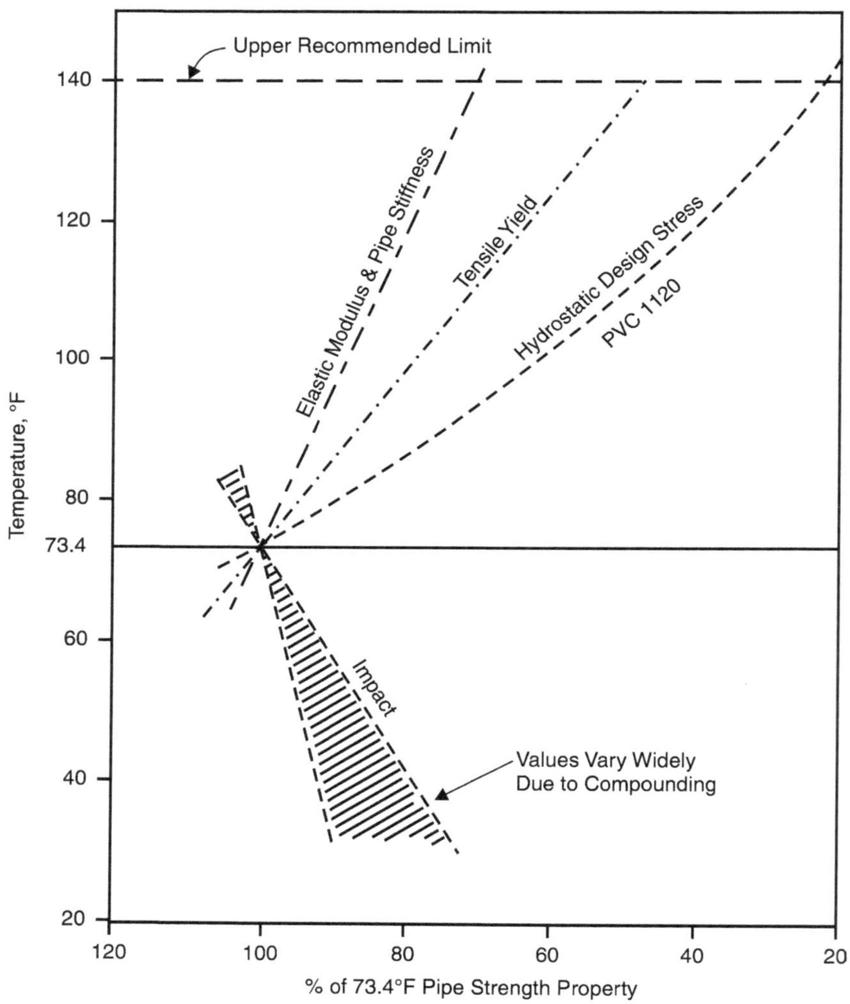

Figure 1-2 Approximate relationship for 12454 PVC for PVC pipe strength properties versus temperature

increases, the impact strength and flexibility of PVC pipe increases. However, with the increase in temperature, PVC pipe loses tensile strength and stiffness; consequently, the pressure capacity of the pipe will be reduced, and more care will be needed during installation to avoid excessive deflection.

Most municipal water systems operate at temperatures at or below 73.4°F (23°C). In these applications, the actual pressure capacity of PVC pipe will be equal to or greater than the product's rated pressure. Intermittent water system temperatures above 73.4°F (23°C) do not warrant de-rating of pipe or fitting pressure designations.

New users and installers of PVC pipe should be aware of the pipe's capacity to expand and contract in response to changes in temperature. The PVC coefficient of thermal expansion is roughly five times the normal value for cast iron or steel. Provisions must be made in design and installation to accommodate expansion and contraction if the pipeline is to provide service over a broad range of operating temperatures. In general, allowance must be made for ⅜ in. (5.4 mm) of expansion or contraction for every 100 ft (30.5 m) of pipe for each 10°F (5.6°C) change in temperature. Gasket joints provide excellent allowance for thermal expansion and contraction of PVC pipelines. The coefficient of thermal expansion for PVCO is the same as for PVC.

Resistance to Biological Attack

PVC pipe is nearly totally resistant to biological attack. Biological attack can be described as degradation or deterioration caused by the action of living microorganisms or macroorganisms. Microorganisms that attack organic materials are normally listed as fungi and bacteria. Macroorganisms that can affect organic materials located underground include an extremely broad category of living organisms; for example, grass roots, termites, and rodents. The performance of PVC pipe in environments providing severe exposure to biological attack in its various anticipated forms has been studied and evaluated since the 1930s.

PVC pipe will not deteriorate or break down under attack from bacteria or other microorganisms, nor will it serve as a nutrient to microorganisms, macroorganisms, or fungi. *The interior walls of PVC pipe or construction debris left in the pipe can support internal growth of biofilms.* No cases have been documented where buried PVC pipe products have degraded or deteriorated because of biological action. As a result, no special engineering or installation procedures are presently required to protect PVC or PVCO pipe from any known form of biological attack.

Elastomeric seals are manufactured from organochemical materials, which can be formulated to produce a variety of properties. Some elastomers are susceptible to biological attack, whereas others provide resistance comparable to those inherent in polyvinyl chloride. PVC pipe manufacturers select gaskets produced from elastomeric compounds that provide high resistance. A material that will not support bacterial growth is a requirement, particularly in potable water systems.

In normal practice, when installing PVC pipe with gasketed joints, assembly of joints is facilitated using a lubricant applied in accordance with the manufacturer's instructions. Care must be exercised in the selection of this lubricant to ensure compatibility with the elastomeric seal and the PVC pipe and to ensure that the lubricant will not support the growth of fungi or bacteria. Care must also be taken to ensure that only the amount of lubricant required to facilitate assembly is used. Excess lubricant can adversely affect water quality and ultimately delay commissioning of a water system. Only the lubricant recommended by the pipe manufacturer should be used. These lubricants must also satisfy all NSF 61 requirements.

Weathering Resistance

PVC pipe can incur surface damage when subjected to long-term exposure to ultraviolet (UV) radiation from sunlight, an effect called ultraviolet degradation. Unless specifically formulated to provide substantial protection from UV radiation, or unless a limited service life is acceptable, PVC pipe is not recommended for applications where it will be continuously exposed to direct sunlight without some form of physical protection (such as paint or wrapping).

Ultraviolet degradation in PVC occurs when energy from the UV radiation causes excitation of the molecular bonds in the plastic. The resulting reaction occurs only on the exposed surface of PVC pipe to a depth of only 0.001–0.003 in. (0.025–0.075 mm). Within the affected zone of reaction, the structure of the PVC molecule is permanently altered with the molecules being converted into a complex structure typified by polyene formations. The polyene molecule causes a light-yellow coloration on the PVC pipe and slightly increases its tensile strength.

Regarding the organochemical reactions that characterize UV deterioration of PVC, the following should be noted:

- Chalking or frosting should not be confused with UV degradation. Chalking is a powdery residue on the surface of a material resulting from degradation and/or migration of an ingredient. Chalking (also known as frosting, haze, or bloom) is a light-scattering surface resembling fine crystals. In these cases, the color change is an aesthetic issue but does not denote UV degradation.
- UV degradation has occurred when the color has changed to yellow, tan, brown, or black.
- UV degradation results in a slight increase in tensile strength, slight increase in the modulus of tensile elasticity, and decrease in impact strength in PVC pipe.
- UV degradation does not continue when exposure to UV radiation is terminated.
- UV degradation occurs only in the plastic material directly exposed to UV radiation and to an extremely shallow penetration depth (the first 0.001–0.003 in. of exposed surface.)
- UV degradation of PVC pipe formulated for buried use will not have significant adverse effect with up to two full years of outdoor weathering and direct exposure to sunlight.

The above is also true regarding PVCO pipe.

Abrasion

After years of investigation and observation, it has been established that the combination of PVC resin, extenders, and various additives in PVC compounds, plus the methods of extrusion for PVC pipe, produce a resilient product with good resistance to abrasive conditions.

Many investigations and tests have been conducted, both in North America and Europe, by manufacturers, independent laboratories, and universities seeking to define PVC pipe's response to abrasion. Although the approaches to the various tests and investigations have varied substantially, the data developed have been consistent in defining the extent of PVC pipe resistance to abrasion. The nature and resiliency of PVC pipe cause it to gradually erode over a broad area when exposed to extreme abrasion rather than to develop the characteristic localized pitting and more rapid failure observed in pipe products with lower abrasion resistance.

PVC pipe is well suited to applications where abrasive conditions are anticipated. In extremely abrasive exposures, wear must be anticipated; however, in many conditions PVC pipe can significantly reduce maintenance costs incurred because of extreme abrasion. It should be noted that potable water, regardless of its makeup, is not considered abrasive to PVC pipe.

In trenchless applications, PVC has continued to exhibit its resistance to external abrasion. As a real-world example, a 10-in. DR 14 PVC pipe was installed under the Beaufort River in South Carolina as part of a 5,120 linear foot horizontal directional drilling installation. The soil conditions included silty sands and some areas of shell deposits. The scratch depths measured in the pipe exterior behind the pull head were ≤2.5% of the pipe wall thickness—after being pulled nearly a mile underground.

Tuberculation

Soluble encrustants (such as calcium carbonate) in some water supplies do not precipitate onto the smooth walls of PVC or PVCO pipe. Because these materials do not corrode, there is no tuberculation caused by corrosion by-products.

REFERENCES

Abrasion Resistance, Das Kunststoffrohr V. 13 (25). (July 1969).

American National Standards Institute/American Water Works Association (ANSI/AWWA). 2016. C900. Polyvinyl Chloride (PVC) Pressure Pipe and Fabricated Fittings, 4 In. Through 60 In. (100 mm Through 1,500 mm). Denver: AWWA.

ANSI/AWWA. 2010. C905. Polyvinyl Chloride (PVC) Pressure Pipe and Fabricated Fittings, 14 In. Through 48 In. (350 mm Through 1,200 mm) for Water Transmission and Distribution. Denver: AWWA.

ANSI/AWWA. 2016. C909. Molecularly Oriented Polyvinyl Chloride (PVCO) Pressure Pipe, 4 In. (100 mm) and Larger. Denver: AWWA.

Arnold, G.E. 1960. Experience With Main Breaks in Four Large Cities: Panel Discussion. *Journal AWWA*, 52:8:1041–1058.

ASTM D1755. 2015. Standard Specification for Poly(Vinyl Chloride) Resins. West Conshohocken, Pa.: ASTM International.

ASTM D1784. 2011. Standard Specification for Rigid Poly(Vinyl Chloride) (PVC) Compounds and Chlorinated Poly(Vinyl Chloride) (CPVC) Compounds. West Conshohocken, Pa.: ASTM International.

ASTM D2241. 2015. Standard Specification for Poly (Vinyl Chloride) (PVC) Plastic Pipe (SDR Series). West Conshohocken, Pa.: ASTM International.

ASTM D2837. 2013. Standard Test Method for Obtaining Hydrostatic Design Basis for Thermoplastic Pipe Materials or Pressure Design Basis for Thermoplastic Pipe Products. Conshohocken, Pa.: ASTM International.

ASTM F477. 2014. Standard Specification for Elastomeric Seals (Gaskets) for Joining Plastic Pipe. West Conshohocken, Pa.: ASTM International.

Babbitt, H.E. 1967 (6th ed.). *Water Supply Engineering*. New York: McGraw-Hill.

Bulkey, C.W.; Morin, R.G.; and Stockwell, A.J. 1968. Vinyl Polymers and Copolymers. *Modern Plastics Encyclopedia*, 45(Oct.):14A:336.

Hendricks, J.C. 1955. Weathering Properties of Vinyl Plastics. *Plastics Technology*.

Hendricks, J.C. and White, E.L. 1952. *Weathering Characteristics of Polyvinyl Chloride Type Plastics*. Brooklyn, N.Y.: National Lead Company Research Laboratories, Wire and Wire Products.

Henson, J.H.L. and Whelan, A. 1973. *Developments in PVC Technology*. London: National College of Rubber Technology.

Hertzberg, L.G. 1956. Suggested Nontechnical Manual on Corrosion for Water Works Operators. *Journal AWWA*, 48(6):719–738.

Modern Plastics Encyclopedia. Issued annually by Modern Plastics. New York:McGraw-Hill.

NSF International. 2014. NSF/ANSI 61: Drinking Water System Components—Health Effects. Ann Arbor, Mich.: NSF International.

Ong, S.K.; Gaunt, J.A.; Mao, F.; and C. Cheng. 2008. *Impact of Hydrocarbons on PE/PVC Pipes and Pipe Gaskets #3946,* Denver: Water Research Foundation.

Penn, W.S. 1967. *PVC Technology*. New York: Wiley Interscience, a Division of John Wiley and Sons Inc.

Plastics Pipe Institute. 2000. TR-19: Thermoplastic Piping for the Transport of Chemicals. Irving, Tex.: Plastics Pipe Institute.

Reedy, D.R. 1966. Corrosion in the Water Works Industry. *Materials Protection*, 5(9):55.

Relative Abrasion Resistance of Ring-Tite PVC Pipe. 1972. Long Beach, Calif.: Johns-Manville Sales Corporation.

Sudrabin, L.P. 1956. Protect Pipes from External Corrosion. *The American City and County*.

Test to Determine Effect of an Undersized (Smaller Diameter Than Inside Diameter of Pipe and Fittings) Electrical Sewer Pipe Auger on Schedule 40 PVC—1 Drain, Waste, and Vent Pipe and Fittings. 1963. Aurora, Ohio: Research Laboratory, Carlon.

Tiedeman, W.D. 1955. A Study of Plastic Pipe for Potable Water Supplies. Ann Arbor, Mich: NSF International.

Tipps, C.W. 1966. Underground Corrosion. *Materials Protection*.

Tobin, W.W. 1965. Stabilization of Rigid Polyvinyl Chloride Against Ultraviolet Radiation. Presented at Society of Plastic Engineers 21st Annual Technical Conf., Boston, Mass.

Transport fester Stoffe durch PVC—hart—Rohre (Transport of Solid Substances Through Hard-PVC Pipes). Code: 237-4032-1, German.

Uni-Bell PVC Pipe Association. 2012 (5th ed.). *Uni-Bell Handbook of PVC Pipe—Design and Construction*. Dallas, Tex.: Uni-Bell PVC Pipe Association.

US Environmental Protection Agency. 1974. *Manual on Sulfides in Sewers*. Washington, DC: USEPA.

The Institute for Hydro-mechanic and Hydraulic Structures. 1973. Wear Data of Different Pipe Materials at Sewer Pipelines. Germany: Technical University of Darmstadt.

Wesfield, L.B.; Thacker, G.A.; and L.I. Nass. 1965. Photodegradation of Rigid Polyvinyl Chloride. *SPE Journal*, 21:7:649.

Wolter, F. 1974. Effect of Outdoor Weathering on the Performance of Some Selected Plastic Piping Materials. Presented by Battelle at the American Gas Association Fifth Plastic Pipe Symposium, Houston, Tex.

The Los Angeles Rubber Group, Inc. 1970. *Yearbook and Directory*. Los Angeles: The Los Angeles Rubber Group, Inc.

Chapter 2

Testing and Inspection

PVC pipe manufacturing has continuously evolved and improved since it began in the 1930s. Today's PVC compounding operations are highly automated and controlled. PVC pipe extrusion, molding, and fabrication operations are technologically sophisticated and closely monitored. This chapter covers testing and inspection as it applies to the manufacturing of PVC and PVCO pipe and their related products.

TESTING AND INSPECTION

Testing and inspection in PVC pipe manufacturing may be divided into three categories: (1) qualification testing, (2) quality control testing, and (3) assurance testing.

Qualification Testing

Qualification testing is performed on piping products and on the materials from which they are produced to ensure that the finished products meet the requirements of applicable specifications. Qualification testing must demonstrate that the materials, process equipment, and manufacturing technology consistently yield, through proper production procedures and controls, finished products that comply with applicable standards.

Material Qualification Requirements

Early in the evolution of the PVC pipe industry, it was recognized that the sustained strength capacity of a PVC compound is significantly influenced by duration of loading. For a more complete discussion of PVC and PVCO pipes' behavior under sustained pressure, please see Chapter 5.

It was also recognized that a PVC compound's long-term capacity to handle sustained stress is not only determined by the base PVC polymer but is also influenced by the nature and quantity of ingredients (such as extrusion processing aids, heat stabilizers, impact modifiers, UV inhibitors, and colorants) that are used to enhance performance, facilitate processing, and provide product identification. Consequently, testing requirements and rules have been developed to ensure that the PVC compounds used to manufacture PVCO and PVC pressure pipe and PVC fittings are capable of providing minimum required stress capacity, over both the short and long term.

The formulation of each PVC compound for pressure pipe is fixed and specific, with the amount and identity of each ingredient spelled out. The ingredient identification includes its chemical and physical properties. When more effective and cost competitive ingredients are incorporated, companies are required to qualify alternative suppliers and products. Early on, a company had to generate extensive long-term stress-rupture data that demonstrated the proposed ingredient supplier change would not compromise the compound's long-term strength. The knowledge gained from qualifying countless substitutions of the basic PVC compound ingredients over the span of several decades led to policies whereby an ingredient can qualify for substitution provided that its physical and chemical properties comply with requirements established for that class of ingredient.

In 1985, a generic and public "PVC range formulation," along with related policies, was adopted. Since that time, many new ingredient suppliers and alternative ingredients have been qualified for use in the PVC range formulation.

The strength qualification requirements for PVC compounds used to manufacture PVC and PVCO pressure pipe and PVC pressure fittings are included in Appendix A. In addition, PVCO and PVC pressure pipe and PVC fittings are required to be made from PVC resin that has been compounded to provide physical and chemical properties that equal or exceed cell class 12454 as defined in ASTM D1784.

Pipe Qualification Requirements

PVC compounds and products to be used for potable water shall be certified by an accredited testing agency for compliance with NSF/ANSI 61.

The following qualification tests are required in the manufacture of ANSI/AWWA C900 PVC and C909 PVCO pipe to evaluate the design properties noted.

Gasket joint design testing. One option for testing joint design is to perform pressure tests to verify that joint assemblies qualify for a hydrostatic design basis category of 4,000 psi (27.6 MPa) for PVC pipe or 7,100 psi (49.0 MPa) for PVCO pipe.

Joint performance testing. All joints must be designed to provide a water tight seal. A representative size of each assembled joint design must meet the requirements of ASTM D3139. Fusion joints shall also provide axial strength for trenchless installation and operational loading. A full wall thickness coupon sample (cut across a joint that has been thermally butt-fused using commercially available equipment in accordance with the manufacturer's recommended procedures and parameters) must withstand a minimum axial stress of 6,400 psi following test procedures set forth in ASTM D638.

Lap-shear test. This test is used to verify that fabricated-fitting solvent cementing procedures result in minimum average lap-strengths of 900 psi (6.2 MPa). Lap-shear test samples are produced by solvent-cementing of component pipe segments identical to those that are used to fabricate fittings.

Quality Control Testing

Quality control testing is routinely performed on specimens of PVC piping products as they are manufactured to ensure that the products comply with applicable standards.

Quality control testing includes, but is not limited to, inspection and testing to verify proper dimensional, physical, and mechanical properties. Frequently, quality control tests are required that may not define a desired finished product property but that do verify the use of proper procedures and controls in the manufacturing process. Quality control tests and inspection required in the manufacture of ANSI/AWWA C900 PVC and C909 PVCO products are as follows.

Workmanship inspection. Inspection is conducted to ensure that the PVC or PVCO pipe product is homogeneous throughout—free from voids, cracks, inclusions, and other defects—and reasonably uniform in color, density, and other physical properties. Surfaces are inspected to ensure that they are free from nicks, gouges, severe scratches, and other such blemishes. Joining surfaces are inspected to ensure freedom from damage and imperfections.

Marking inspection. Inspection verifies proper marking of the pipe as required in the applicable product standard. Marking of ANSI/AWWA C900 PVC and C909 PVCO pipe includes the following:

- PVC or PVCO
- Manufacturer's name or trademark and production-record code
- Nominal pipe size
- Outside diameter (C900 only)
- Dimension ratio (DR; e.g., DR 25)
- AWWA pressure class (e.g., PC 100)
- AWWA standard designation (e.g., ANSI/AWWA C900)
- Seal of the testing agency that verified the suitability of the pipe material for potable-water service
- In compliance with NSF/ANSI 61 or, if not intended for potable water, the words "NOT FOR POTABLE USE"
- For deflectable joints, the maximum allowable axial joint deflection in degrees (e.g., joint D ≤ 1°)

Markings of machined couplings and fabricated fitting include the following:

- Nominal size and deflection angle (if applicable)
- AWWA pressure class
- AWWA standard designation
- Manufacturer's name or trademark
- Seal of the testing agency that verified the suitability of the pipe material for potable-water service in compliance with NSF/ANSI 61 or, if not intended for potable water, the words "NOT FOR POTABLE USE"
- For deflectable joints, the maximum allowable axial joint deflection in degrees (e.g., joint D ≤ 1°)

Dimension measurement. Measurement of dimensions on a regular and systematic basis is essential. Failure to meet dimensional requirements may render the product unsatisfactory regardless of success in other inspections and tests. All dimensional measurements are made in accordance with ASTM D2122 and include the following:

- Product diameter
- Product wall thickness
- Bell joint dimensions
- Fabricated-fitting configurations
- Length

Some dimensional requirements are defined in the manufacturer's product specifications.

Product packaging inspection. The finished package of PVC pipe prepared for shipment to the customer is inspected to ensure correct pipe quantity and adequate protection of the pipe.

Quick-burst test. The PVC pipe sample is pressurized to burst within the test time period of 60–70 s. Burst pressure measured must not be less than the minimum burst pressure requirements defined in ANSI/AWWA C900 or C909. Quick-burst testing is conducted in accordance with ASTM D1599. ASTM D2290 may be used in place of quick-burst testing of pipe. This test is also performed on machined couplings.

Flattening test. The PVC pipe specimen is partially flattened between moving parallel plates. When the pipe is flattened 60 percent (the distance between the parallel plates equals 40 percent of the original outside diameter), the specimen must display no evidence of splitting, cracking, or breaking.

Extrusion quality test. The PVC pipe specimen is immersed in anhydrous (dry) acetone for 20 min. When removed from the acetone bath, the pipe specimen must pass the failure criteria in ASTM D2152. Extrusion quality testing is conducted in accordance with ASTM D2152 and distinguishes only between unfused and properly fused PVC pipe.

Quality control inspection and testing must not be confused with field acceptance testing. Quality control testing is appropriate only during or immediately following the manufacturing process.

Arc test for fabricated fittings. The arc test is required for butt-fused or thermally welded joints in fabricated fittings. Any discontinuity in a segment joint is indicated by the presence of an arc (spark) from a probe tip and is cause for rejection of the fitting.

Fabricated-fitting pressure test. In this test, the fabricated fitting must not fail, balloon, burst, or weep when subjected to an internal pressure test. For ANSI/AWWA C900 fabricated fittings, the internal pressure test is equal to two times its designated pressure class for a minimum of two hours.

Assurance Testing

Assurance testing is performed at the completion of the manufacturing process to assure the finished products consistently and reliably satisfy the requirements of applicable standards. Quality assurance tests required in the manufacture of ANSI/AWWA C900 PVC and C909 PVCO products are as follows.

Sustained pressure test. ANSI/AWWA C909 pipe and C900 pipe or fabricated fittings shall not fail, balloon, burst, or weep as defined in ASTM D1598 at the applicable sustained pressure when tested for 1,000 hr.

Hydrostatic proof test. The hydrostatic proof test is required in the manufacture of PVC and PVCO pipe as well as machined PVC couplings in accordance with ANSI/AWWA C900 and C909. In the test, every PVC coupling and every standard length of PVC or PVCO pipe is proof-tested for a minimum dwell time of 5 s. The standards require that the hydrostatic proof test pressure be a minimum of two times the pressure class of the pipe (i.e., 2 × 235 psi = 470 psi for PC235 pipe).

The hydrostatic proof-test frequency for nonstandard lengths of pipe is every 24 hrs. The hydrostatic proof-test frequency may also be modified by agreement between manufacturer and producer/supplier.

REFERENCES

American National Standards Institute/American Water Works Association (ANSI/AWWA). 2016. C900. Polyvinyl Chloride (PVC) Pressure Pipe and Fabricated Fittings, 4 In. Through 60 In. (100 mm Through 1,500 mm). Denver: AWWA.

ANSI/AWWA. 2016. C909. Molecularly Oriented Polyvinyl Chloride (PVCO) Pressure Pipe, 4 In. (100 mm) and Larger. Denver: AWWA.

ASTM D1598. 2015. Standard Test Method for Time-to-Failure of Plastic Pipe Under Constant Internal Pressure. West Conshohocken, Pa.: ASTM International.

ASTM D1599. 2018. Standard Test Method for Resistance to Short-Time Hydraulic Pressure of Plastic Pipe, Tubing, and Fittings. West Conshohocken, Pa.: ASTM International.

ASTM D1784. 2011. Standard Specification for Rigid Poly(Vinyl Chloride) (PVC) Compounds and Chlorinated Poly(Vinyl Chloride) (CPVC) Compounds. West Conshohocken, Pa.: ASTM International.

ASTM D2122. 2016. Standard Test Method for Determining Dimensions of Thermoplastic Pipe and Fittings. West Conshohocken, Pa.: ASTM International.

ASTM D2152. 2017. Standard Test Method for Adequacy of Fusion of Extruded Poly(Vinyl Chloride) Pipe and Molded Fittings by Acetone Immersion. West Conshohocken, Pa.: ASTM International.

ASTM D2290. 2019. Standard Test Method for Apparent Hoop Tensile Strength of Plastic or Reinforced Plastic Pipe. West Conshohocken, Pa.: ASTM International.

Heilmayr, P.F. 1976. PVC Pipe Keeps Rolling Along. *Plastics Engineering*.

NSF International. 2014. NSF/ANSI 61: Drinking Water System Components—Health Effects. Ann Arbor, Mich.: NSF International.

Sarvetnick, H.A. 1969. *Polyvinyl Chloride*. New York: Van Nostrand Reinhold Co.

Winding, C.C. and Hiatt, G.D. 1961. *Polymeric Materials*. New York: McGraw-Hill.

Uni-Bell PVC Pipe Association. 2012 (5th ed.). *Uni-Bell Handbook of PVC Pipe—Design and Construction*. Dallas, Tex.: Uni-Bell PVC Pipe Association.

This page intentionally blank.

Chapter 3

Hydraulics

Many empirical formulas and equations have been developed to provide a solution to the problem of flow in pipes and are used daily by water utility engineers. Relatively few specific problems in pipe hydraulics, such as laminar flow, can be solved entirely by rational mathematical means. Solutions to the majority of flow problems depend on experimentally determined coefficients and relationships. Commonly used flow formulas have been developed through research by Darcy, Chezy, Kutter, Scobey, Manning, Weisbach, Hazen, and Williams.

FLOW FORMULAS

Hydraulic flow research and analysis have established that the Hazen–Williams equation can be used for PVC pressure piping system design. Flow conditions may also be analyzed more precisely and with more detail using the Darcy–Weisbach equation.

Darcy–Weisbach Equation

The Darcy–Weisbach equation provides the hydraulic design of PVC pressure water pipe. Relative pipe roughness (ε/D) and Reynolds number ($R_e = VD/v$) are also defined. The commonly used form of the Darcy–Weisbach formula is shown in Eq 3-1.

$$h_f = f \frac{LV_f^{\,2}}{D2g} \quad \text{(Eq 3-1)}$$

where:

h_f = head loss, ft of H_2O
f = friction factor
L = pipe length, ft
D = pipe inside diameter, ft
V_f = mean flow velocity, ft/s
g = acceleration of gravity, 32.2 ft/s^2

Investigation and analysis by Neale and Price have established that the friction factor f for a hydraulically smooth flow in PVC pipe may be defined by the following equation:

$$\frac{1}{\sqrt{f}} = 2\log_{10}\left(R_e\sqrt{f}\right) - 0.8 \qquad \text{(Eq 3-2)}$$

where:

f = friction factor
R_e = Reynolds number

The calculations for the friction factor (f) may be tedious. In common practice, the factor is established by using the Moody diagram shown in Figure 3-1. Relative roughness (ε/D) is related to friction factor (f) as shown in Eq 3-3. Figure 3-2 provides values for relative roughness (ε/D) for various pipe products.

$$\frac{1}{\sqrt{f}} = 1.14 - 2\log_{10}\left(\frac{\varepsilon}{D} + \frac{9.35}{R_e\sqrt{f}}\right) \qquad \text{(Eq 3-3)}$$

where:

f = friction factor
ε = 0.000005 ft, PVC pipe
D = pipe inside diameter, ft
R_e = Reynolds Number

Hazen–Williams Equation

The Hazen–Williams flow equation is the most widely accepted and used for calculating pressure pipe flow conditions. The equation can be expressed in the following ways depending on the solution needed. Flow velocity in a pipeline can be calculated using Eq 3-4.

$$V = 1.318C\left(R_H\right)^{0.63}\left(S\right)^{0.54} \qquad \text{(Eq 3-4)}$$

where:

V = flow velocity, ft/s
C = flow coefficient
R_H = hydraulic radius, ft
Note: $R_H = \frac{1}{4}(D)$ for pipe flowing full
S = hydraulic slope, ft/ft

Flow rate in gpm, given pressure drop in psi, can be calculated using Eq 3-5.

$$Q = 0.442 d_i^{2.63} C\left(\frac{P_1 - P_2}{L}\right)^{0.54} \qquad \text{(Eq 3-5)}$$

where:

Q = flow rate, gpm (All gallons are US gallons unless otherwise noted.)
d_i = pipe inside diameter, in.
C = flow coefficient
P_1, P_2 = gauge pressures, psi
L = pipe length, ft

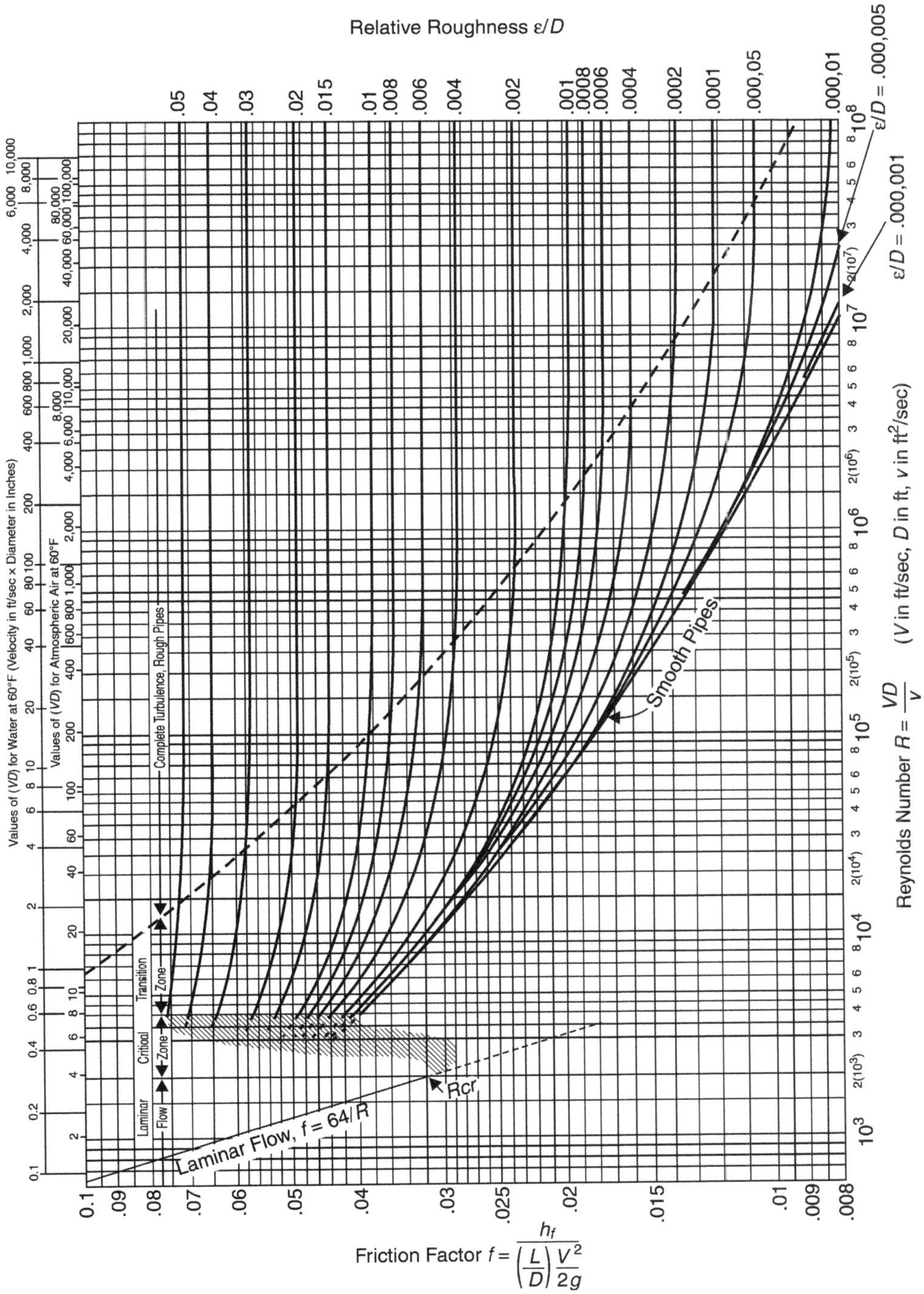

Source: American Society of Mechanical Engineers, New York, Transactions, ASME, Vol. 66 (1944) L F. Moody.

Figure 3-1 Moody diagram—friction factor

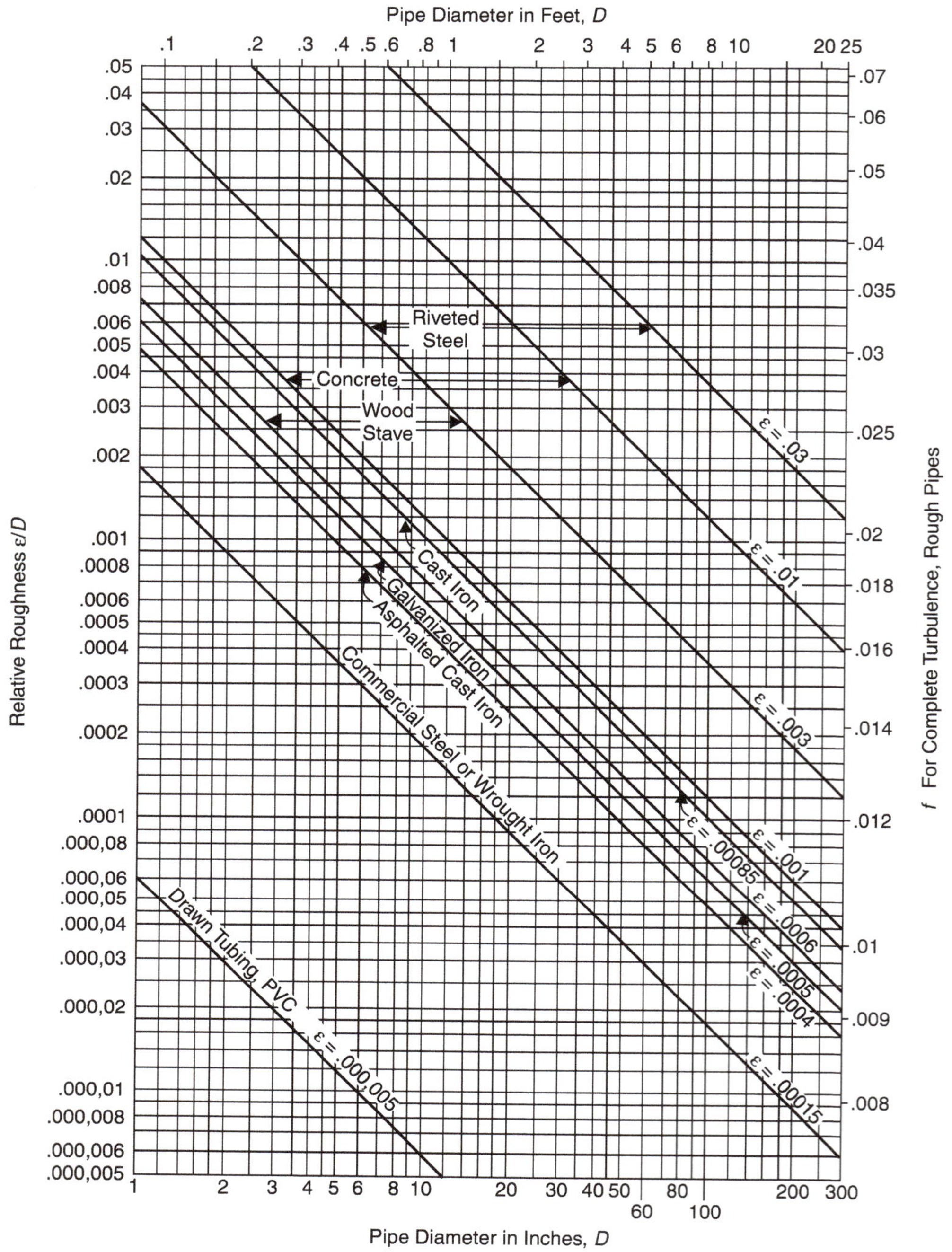

Source: American Society of Mechanical Engineers, New York, Transactions, ASME, Vol. 66 (1944) L.F. Moody.

Figure 3-2 Moody diagram—relative roughness, where ε is in ft

Using Eq 3-6, flow rate can be derived from pressure drop expressed in terms of feet per 1,000 ft.

$$Q = 0.006756\, C d_i^{2.63}\, H^{0.54} \tag{Eq 3-6}$$

where:

Q = flow rate, gpm
C = flow coefficient

d_i = pipe inside diameter, in.
H = head loss, ft/1,000 ft

Friction loss in hydraulic flow can be derived from Eq 3-7.

$$f = 0.2083\left(\frac{100}{C}\right)^{1.852}\frac{Q^{1.852}}{d_i^{\,4.8655}} \qquad \text{(Eq 3-7)}$$

where:

f = friction loss, ft of water/100 ft
C = flow coefficient
Q = flowrate, gpm
d_i = pipe inside diameter, in.

Flow coefficients for PVC pipe have been derived through research and analysis by various researchers, including Neale, Price, Bishop, and Jeppson. Research has established that the Hazen–Williams flow coefficient C can range in value from 155 to 165 for both new and previously used PVC pipe. Therefore, a flow coefficient of $C = 150$ is generally used as a conservative value for determining the headloss through PVC pipe. The roughness value may decrease after the pipeline is in service if there is scale or biofilm buildup. While a Hazen–William's flow coefficient of 150 is considered conservative for calculating steady-state hydraulics on new PVC pipe, a lower coefficient should be considered to account for minor losses if not otherwise specifically accounted for (valves, bends, etc.).

Using $C = 150$ for PVC pipe, Eq 3-4 through 3-7 can be simplified as follows:

$$V = 197.7\,R_H^{\,0.63}\,(S)^{0.54} \qquad \text{(Eq 3-8)}$$

$$Q = 66.3d_i^{\,2.63}\left(\frac{P_1 - P_2}{L}\right)^{0.54} \qquad \text{(Eq 3-9)}$$

$$Q = 1.0134\,d_i^{\,2.63}\,H^{0.54} \qquad \text{(Eq 3-10)}$$

$$f = 0.0984\frac{Q^{1.85}}{D_i^{\,4.86}} \qquad \text{(Eq 3-11)}$$

where:

V = flow velocity, ft/s
R_H = hydraulic radius, ft
S = hydraulic slope, ft/ft
Q = flowrate, gpm
d_i = pipe inside diameter, in.
P_1, P_2 = gauge pressures, psi
L = pipe length, ft
H = head loss, ft/1,000 ft
f = friction loss, ft of water/100 ft

For convenience in design, Tables B-1, B-2, and B-3 in Appendix B have been developed based on the Hazen–Williams formula with $C = 150$ to provide flow capacity (gpm), friction loss (feet per 100 ft), and flow velocity (ft/s) for ANSI/AWWA C900, ASTM D2241, and C909 PVC pressure pipe products. Nomographs for solving head loss characteristics are provided in Figures 3-3 and 3-4.

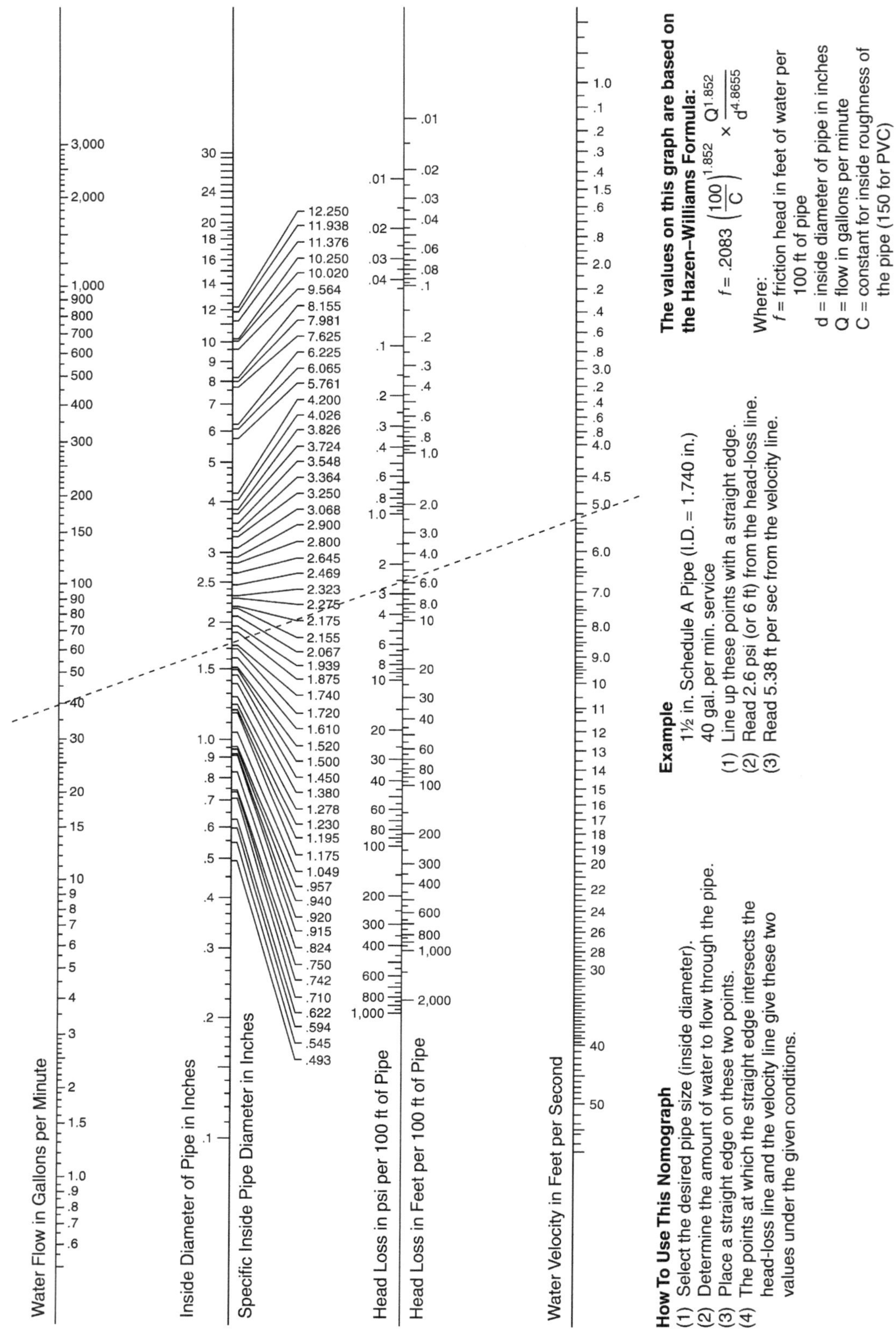

Figure 3-3 Friction loss characteristics of water flow through PVC pipe

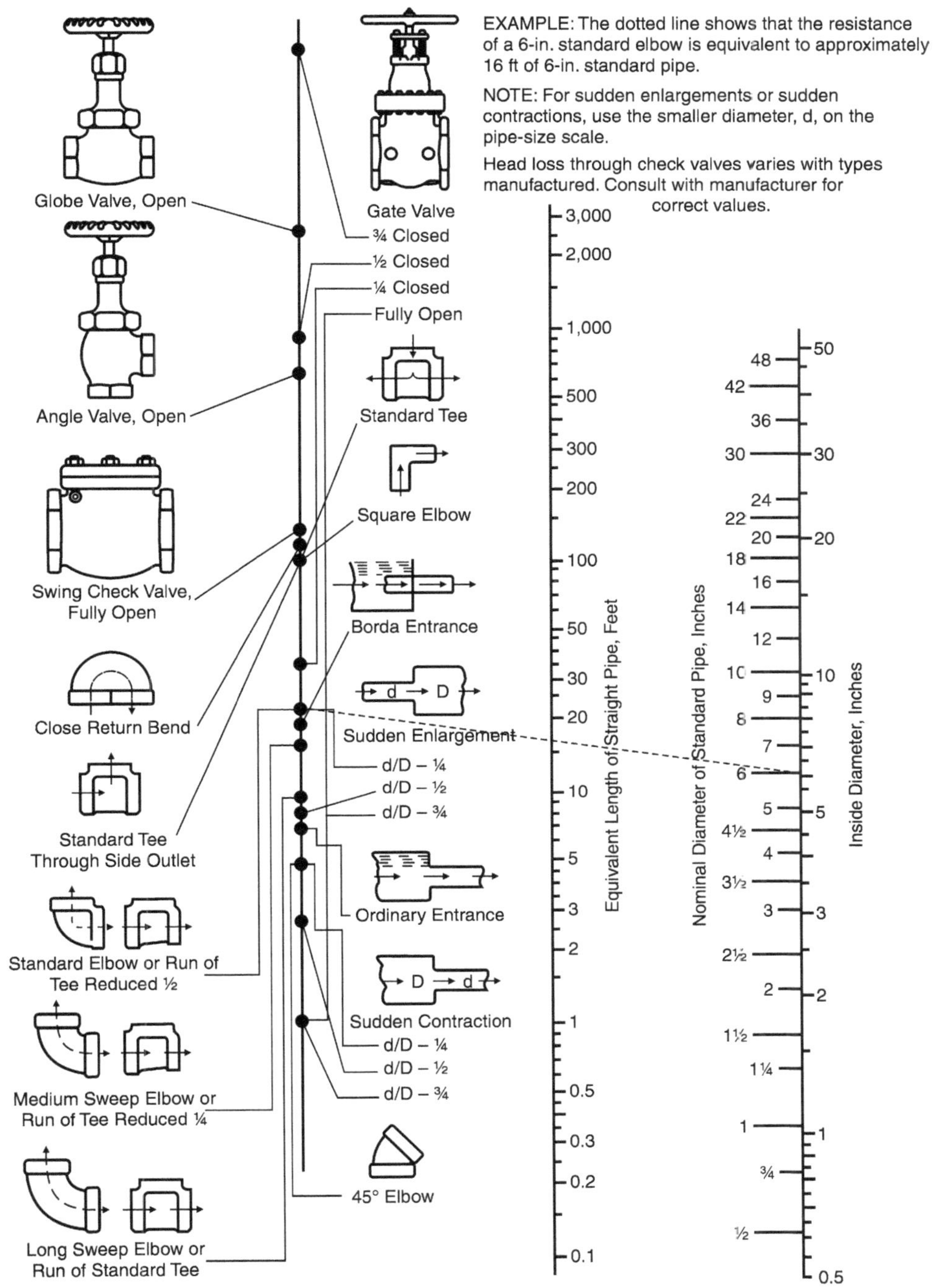

Source: Flow of Fluids through Valves, Fittings, and Pipe. Copyright 1942 by Crane Company.

Figure 3-4 Resistance of valves and fitting to flow of fluids

Impact of Internal Bead on Flow in Fused PVC

The resulting joint of fused PVC contains an internal ridge. This ridge or internal bead has negligible effect on flow or head loss in pressure applications. Using a C value of 150 is recommended for fused PVC, as it is for bell-and-spigot joint PVC and covers the effect of the internal bead as it does for the bell-and-spigot joint. As a result, there is no need to remove the bead in pressure applications.

It should be noted in gravity flow applications that the internal bead can be removed to eliminate a place where a biological buildup could occur.

REFERENCES

American National Standards Institute/American Water Works Association (ANSI/AWWA). 2014. C651. Standard for Disinfecting Water Mains. Denver: AWWA.

ANSI/AWWA. 2016. C900. Polyvinyl Chloride (PVC) Pressure Pipe and Fabricated Fittings, 4 In. Through 60 In. (100 mm Through 1,500 mm). Denver: AWWA.

ANSI/AWWA. 2015. C909. Molecularly Oriented Polyvinyl Chloride (PVCO) Pressure Pipe, 4 In. (100 mm) and Larger. Denver: AWWA.

ASTM D1598. 2015. Standard Test Method for Time-to-Failure of Plastic Pipe Under Constant Internal Pressure. West Conshohocken, Pa.: ASTM International.

ASTM D2241. 2015. Standard Specification for Poly(Vinyl Chloride) (PVC) Plastic Pipe (SDR Series). West Conshohocken, Pa.: ASTM International.

Bishop, R.R. 1975. Hydraulic Characteristics of PVC Pipe in Sanitary Sewers. Logan, Utah: Utah State University. https://digitalcommons.usu.edu/cgi/viewcontent.cgi?article=1597&context=water_rep (accessed Oct. 11, 2019).

Connell, D. Hazen-Williams C-factor Assessment in an Operational Irrigation Pipeline. Montreal, Quebec, Canada: McGill University Department of Agricultural and Biosystems Engineering.

Crane Co. Flow of Fluids Through Valves, Fittings and Pipe. 2011. Technical Paper No. 410.

Hucks, R.T. 1972. Designing PVC Pipe for Water-Distribution Systems. *Journal AWWA*, 64:7:443.

Hucks, R.T. 1972. Design of PVC Water Distribution Pipe. *Civil Engineering*, ASCE, 42:6:70.

Internal Fusion Bead Hydraulics, Technical Bulletin 8-546, Underground Solutions, Inc., Poway, CA, Rev 4 2017-1016.

Neale, L.C. and Price, R.E. 1964. Flow Characteristics of PVC Sewer Pipe. ASCE, *Journal of Sanitary Engineers Div. Proc.* 90 SA3, 109.

Pipe Friction Manual. 1961 (3rd ed.). New York: Hydraulic Institute.

Streeter, V.L. 1958 (3rd ed.). *Fluid Mechanics*. New York: McGraw-Hill.

Uni-Bell Handbook of PVC Pipe—Design and Construction. 2012 (5th ed.). Dallas: Uni-Bell PVC Pipe Association.

Chapter 4

External Load Design for Trench Installation

This chapter presents design considerations for selecting the pipe stiffness, soil properties, and installation requirements to support the earth loads, live loads, and surcharge loads. These parameters may be adjusted to determine an optimum design.

This discussion is for pipe in a trench installation. Trenchless applications, such as pipe bursting and sliplining, where the original host pipe is in place, can use this analysis as well. Horizontal directional drilling (HDD) may also use this method but may require an arched or bridged analysis due to depth in undisturbed soils. As not all soils bridge or arch, specific geotechnical expertise may be required for trenchless HDD installations.

FLEXIBLE PIPE DESIGN

As illustrated in Figure 4-1, PVC and PVCO pipe can deflect due to load, thereby developing passive soil support at the sides of the pipe. At the same time, soil arching over the pipe due to the deflection transfers a portion of the vertical soil load to the soil at the side of the pipe. When designed and installed correctly, the pipe-soil system can effectively support the load on the pipe.

The deflection of the pipe is the change in vertical diameter divided by the original pipe diameter, most often stated as a percentage. Conceptually and without regard for any internal pressure, the relationship between load and deflection may be expressed as shown in Eq 4-1.

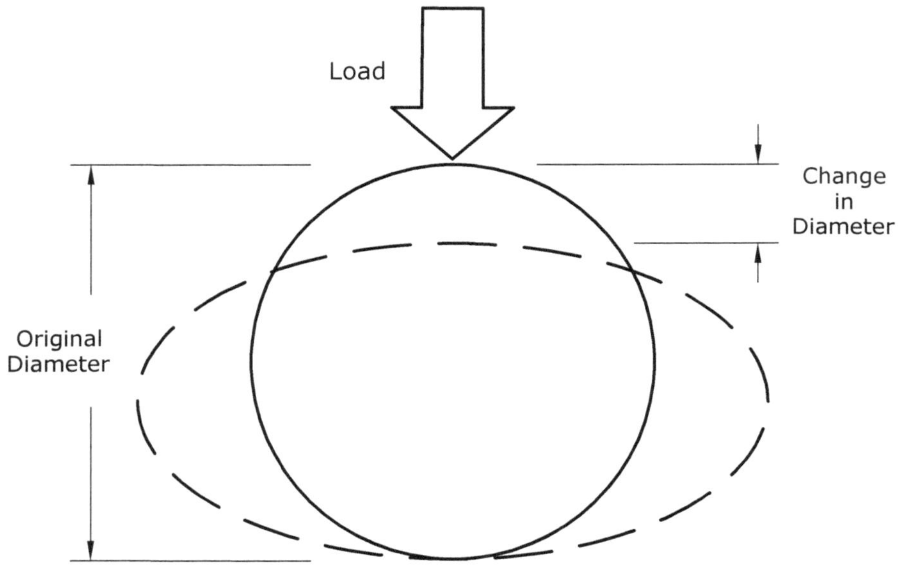

Figure 4-1 Flexible pipe deflection

$$\text{Deflection} = \frac{\text{Load}}{\text{Pipe stiffness} + \text{Soil stiffness}} \quad \text{(Eq 4-1)}$$

The soil stiffness depends on the soil classification and compaction. Because internal pressure usually determines the wall thickness required, PVC and PVCO pressure pipes experience minimal deflection under most soil conditions. The horizontal and vertical deflections may be considered equal in the range of deflections allowed in the AWWA PVC and PVCO pipe standards.

The design methods in this chapter will use specific terminology for the trench construction, soil classification, and soil compaction.

TRENCH TERMINOLOGY

Embedment and backfill materials that surround a buried pipe are defined by their function and location. Figure 4-2 represents the trench terms based on ASTM D2774. This chapter provides information to aid selection of the appropriate embedment soil and the percent compaction required. The methods for proper compaction and subsequent testing are discussed in Chapter 7, "Installation in Trenches and Embankments."

Foundation: The foundation is the native soil at the bottom of the excavation. If the foundation is unsuitable, remediation will be required to provide a stable trench bottom.

Bedding: The bedding is the soil placed at the bottom of the trench on top of the foundation. The bedding serves as a cushion for the pipe.

Haunch Zone: The haunch zone is from the bottom of the pipe up to the springline. The haunch zone and the initial backfill provide side support for flexible pipe that limits deflection. (The springline is the horizontal centerline.)

Initial Backfill: The initial backfill extends from the top of the haunch zone to 12 in. above the top of the pipe. The lower portions of initial backfill (i.e., below the pipe crown), combined with the haunch zone, provide lateral support for flexible pipe.

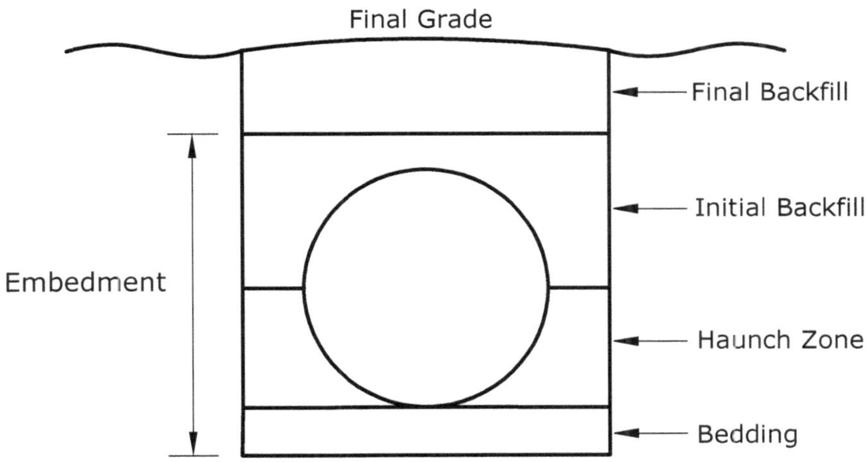

Figure 4-2 Trench construction and terminology

Embedment: The embedment includes the bedding, haunch zone, and initial backfill.

Final Backfill: The final backfill extends from the top of the initial backfill to the final grade.

The haunch zone and initial backfill, from the bottom of the pipe to 75 percent of the outside diameter of the pipe, provide support that limits pipe deflection. The initial backfill above that level serves as a padding over the top of the pipe and reduces the impact of the final backfill when it is dumped into the trench. The maximum particle size in the initial backfill should be specified to avoid pipe damage.

As discussed in Chapter 7 for a wide trench or a trench with sloped walls, the initial backfill padding need not extend across the entire width of the trench and can be about 12 in. thick over the top of the pipe. This reduces the amount of imported soil, if needed, for the initial backfill.

SOIL CLASSES

Table 4-1 gives the soil classes used in ASTM D 2774 to describe the allowable materials for each area around the pipe. The soil classes, Class I to Class V as defined by the Unified Soil Classification System (USCS), are in descending order of stiffness when the soil is compacted. Class I and Class II soils are usually considered cohesionless and are best compacted using vibration. Class III and Class IV are usually considered cohesive and are best compacted with pressure, impact, or kneading. Class V soils are considered cohesive and are not recommended to use as embedment for pipe with low stiffness (high dimension ratio) values.

As discussed in Chapter 7, there is a maximum particle size limit associated with the soil that applies within each of the soil classes.

SOIL COMPACTION

The soil support for the pipe is dependent on the degree of compaction, referred to as *percent compaction*. Percent compaction is defined by ASTM D653 as the ratio of the field compaction to the laboratory maximum density, expressed as a percentage. The field

Table 4-1 Soil classes for pipe installation

Soil Class	Description	USCS Symbol
Class I	Crushed rock 100% passing $1\frac{1}{2}$-in. sieve, ≤25% passing $\frac{3}{8}$-in. sieve, ≤15% passing #4 sieve, ≤12% fines	
Class II	Clean, coarse-grained soils or any soil beginning with one of these symbols (can contain up to 12% fines) (Note 1)	GW GP SW SP
Class III	Coarse-grained soils with fines	GM GC SM SC
	Sandy or gravelly fine-grained soils with ≥30% retained on #200 sieve	ML CL
Class IV	Fine-grained soils with <30% retained on #200 sieve	ML CL
Class V	Fine-grained soils, organic soils, high compressibility silts and clays, organic soil	MH CH, OL OH Pt

Notes:
1. Uniform fine sands (SP) (SP-SC) (SP-SM) with more than 50 percent passing a #100 sieve should be treated as Class III material.
2. Soil classification in accordance with ASTM D2487 or ASTM D2488.
3. *Fines* are soil particles that pass a #200 sieve.
4. Class I: crushed rock particles should have all fractured faces.
5. Recycled concrete, slag, and shells should be considered Class II.

compaction is measured by in-place density tests such as sand cone or nuclear gauge. For soil Classes III, IV, or V, the laboratory maximum density is determined using the standard Proctor compaction test ASTM D698. For soil Classes I or II, the laboratory maximum density is determined using a vibratory compaction test ASTM D7382 or D4253. ASTM D7382 is a new procedure using a vibratory hammer to obtain a maximum density and is considered more reliable than D4253.

References to the percent compaction of soils in this manual will be as recommended in ASTM D653. The percent of the maximum density of the soil is followed by the ASTM test procedure used to determine the maximum density. For example, 95 percent (D698) means that the in-place density should be equal to or higher than 95 percent of the maximum density obtained using D698. The references to soil stiffness in this manual are based on standard Proctor density, which is ASTM D698. Therefore, plans and specifications should reference D698.

Chapter 7 discusses the various methods of compacting soils, measuring densities, and evaluating the percent compaction. In some situations, installing the pipe with the proper soil and compaction may be critical to establishing the correct support for the pipe. Inspection during construction will ensure that the installation meets the design criteria.

In many cases, flowable fill can be substituted for the bedding and embedment. Chapter 7 discusses the proper use of flowable fill for buried pipe.

BASIC AND ENGINEERED INSTALLATIONS

For flexible pipe, there are some combinations of pipe selection, external loads, and soil stiffness that may not need design verification for deflection. Accordingly, the design and construction may be divided into *basic installations* and *engineered installations*.

Basic Installation

Some pressure pipe will have adequate stiffness to withstand the external loads and will not be reliant on the embedment for additional support. Such pipe does not require deflection calculations and may be installed with minimum soil support. The embedment material can be the soil excavated from the trench. The pipe can be laid directly on the trench bottom, and minimal testing and inspection are required.

A basic installation can be used when all of the following conditions are met:

- Nominal pipe size of 24 in. or less.
- DRs equal to or less than 25.
- Depths of cover 15 ft or less.
- Natural ground water below pipe.
- For depths of cover with less than 6 ft, no live load or surcharge load.
- Final backfill compaction is not required.
- Embedment soil E′ will be at least 200 psi (as discussed later).
- Trench foundation and trench walls are stable and have a minimum unconfined compressive strength of 5 psi, an N value of at least 5 from the standard penetration test, or an E′ of at least 400 psi (as discussed later).
- The foundation does not consist of expansive clays, collapsing soils, or landfill.
- The soils in the foundation and used for the embedment do not contain rock particles larger than the maximum particle size discussed in Chapter 7.

An engineered installation should be used when these conditions are not met. In some cases where live or surcharge loads may occasionally occur, such as road crossings, the pipeline system may consist of both basic installations and engineered installations.

Class I, II, III, or IV soils, whether native or imported, can provide an embedment $E' \geq 200$ psi when dumped in place beside the pipe without any compaction. Class V soil may not be a suitable embedment for pipes with low stiffness (high DR) values.

Engineered Installation

When the basic installation is deemed to be not adequate, the pipe design will warrant estimating deflection, as outlined in this chapter. An engineered installation design will need to consider the trench wall support, the effects of ground water, selection of embedment material, increased percent compaction, time before pipeline is pressurized, live load, and surcharge load. Construction may require imported embedment material, placing a bedding for the pipe, soil testing requirements, and more stringent inspection. Where the pipeline crosses under another pipeline, a heavily travelled roadway, or a large waterway, an engineered installation design should be used.

PIPE DEFLECTION

For PVC and PVCO pressure pipe, deflection is usually not a critical design parameter because pipe stiffness values are relatively high and operating internal pressures are usually sufficient to induce re-rounding of the pipe. However, for the *engineered installation*, the deflection should be estimated for those instances when the pipe is not pressurized. A 7.5 percent deflection limit (which is based on a 4 to 1 safety factor over the empirically tested maximum of 30 percent) has been established as a long-term deflection design limit. For the initial deflection, a 5 percent limit should be used.

Deflection Calculation

Pipe deflection will depend upon the pipe's stiffness, the external load(s), and the stiffness of the embedment soil along the sides of the pipe. Deflection can be estimated using Eq 4-2.

$$\%\frac{\Delta Y}{D} = \frac{k(T_L W_E + W_L + W_S)}{0149\text{PS} + 0.061E'}(100) \qquad \text{(Eq 4-2)}$$

where:

$\%\frac{\Delta Y}{D_o}$	=	vertical deflection, %	
ΔY	=	vertical deflection or change in diameter, in.	
D	=	diameter, in.	
k	=	bedding constant, use 0.1	
T_L	=	time-lag factor, use 1.0, 1.5, or 2.0	See Discussion
W_E	=	earth load pressure, psi	See Eq 4-5
W_L	=	live load pressure, psi	See Eq 4-7
W_S	=	surcharge pressure, psi	See Eq 4-6
PS	=	pipe stiffness, lbf/in./in.	See Table 4-3
E'	=	modulus of soil reaction, psi	See Eq 4-14

LOAD ON PIPE

The load on the pipe is the earth load, W_E, plus the live load, W_L, and surcharge load, W_S, if any.

Earth Load

For simplicity, the earth load on the pipe is considered to be a prism load. As indicated in Eq 4-3, the prism load is the weight of the full column height of soil directly over the pipe. Equation 4-4 results in soil pressure in psi for the prism load.

Prism Load: $W_p = Hw\,D$ (lb/lin ft) (Eq 4-3)

Prism load may also be expressed in terms of soil pressure as follows:

Soil Pressure: $P = wH$ (lb/ft^2) (Eq 4-4)

where:

P = pressure caused by soil weight at depth H, lb/ft^2
w = unit weight of soil, lb/ft^3
H = depth at which soil pressure is desired, ft
D = pipe diameter

The prism load can be converted to lb/in^2 for W_E as follows:

Earth Load Pressure: $W_E = \frac{W_P}{144}$ (Eq 4-5)

Surcharge Load

Building foundations, other structure foundations, or other static, long-term loads above the pipe must be considered and are treated as surcharge loads. These loads may be present at the time that the pipe is installed or may be superimposed on the pipe at some point in time subsequent to the pipe installation. These loads increase soil pressure onto the buried pipe and can generally be categorized into two types: (1) loads that have confined footprints or areas of influence, or (2) loads that have wide areas of influence, normally parallel the pipe. Type 1 loads are usually analyzed as point source loads, while type 2 loads are analyzed as uniformly distributed loads.

The basis for analysis of both types of static loads is the Boussinesq theory, which is mathematically stated as

$$W_S = P_A = \frac{WZ^3}{96\pi R^5} \qquad \text{(Eq 4-6)}$$

where:

P_A = soil pressure at point A, psi
W = superimposed load, lb
Z = vertical distance from the point of the load to the top of the pipe, ft
R = $\sqrt{X^2 + Y^2 + Z^2}$, straight line distance from the point where the load is applied to point A on the top of the pipe; X and Y are measured horizontally at 90 degrees to each other. (*Note:* Y is usually measured along the horizontal axis of the pipe, and X is usually measured perpendicular to the horizontal axis of the pipe.)

Live Loads

The following calculations can be used to compute the live load on the pipe for surface traffic (see Figure 4-3). The procedure is based on the AASHTO LRFD (American Association of State Highway and Transportation Officials load-and-resistance factor design philosophy) Bridge Design Specification, 6th ed., 2012. These calculations are for a single-axle truck traveling perpendicular to the pipe on an unpaved surface or a road with flexible pavement. With the inclusion of the multiple presence factor (M_P), they will generally yield acceptably conservative load estimates. The M_P accounts for the probability relationship between the number of loaded lanes and the weight as any single vehicle. M_P is typically taken as 1.2 for single passing vehicle.

$$W_L = \frac{M_P P I_F}{(L_1)(L_2)} \qquad \text{(Eq 4-7)}$$

where:

W_L = live load on pipe, psi
M_P = multiple presence factor = 1.2
P = wheel load magnitude
= 16,000 lb for AASHTO HS20 truck
= 20,000 lb for AASHTO HS25 truck
I_F = impact factor
L_1 = load width parallel to direction of travel, in.
L_2 = load width perpendicular to direction of travel, in.

$$I_F = 1 + 0.33\frac{(96 - h)}{96} \geq 1.0 I_F \qquad \text{(Eq 4-8)}$$

where:

h = burial depth to top of pipe, in.

$$L_1 = t_\ell + \text{LLDF}\,(h) \qquad \text{(Eq 4-9)}$$

where:

t_ℓ = length of tire footprint = 10 in.
LLDF = factor to account for live load distribution with depth of fill

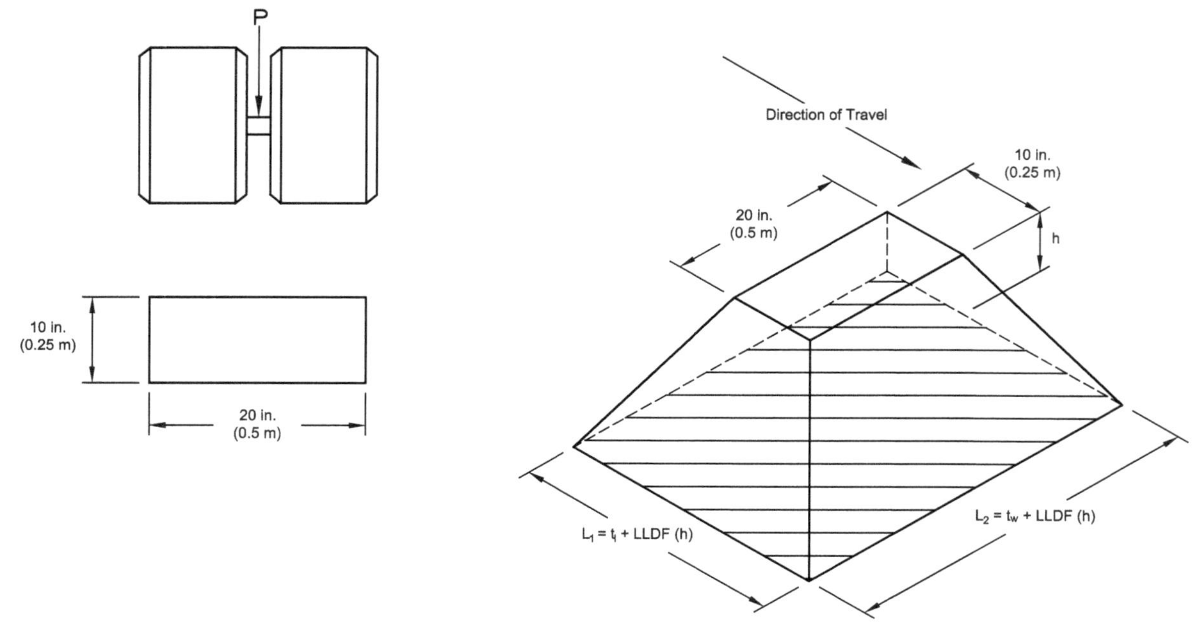

Source: Adapted from AASHTO
Note: For H > 45 in. (1.14 m), see part 2 of the L_2 formula. (Change accounts for overlapping influence areas from adjacent wheel loads.)

Figure 4-3 Distribution of AASHTO HS20 or HS25 live load through granular fill for $h \leq 45$ in. (3.75 ft)

= 1.15 for Class I and Class II backfills
= 1.0 for all other backfills

If $h \leq h_{int}$

$$L_2 = t_w + \text{LLDF}\,(h) \qquad \text{(Eq 4-10)}$$

where:

t_w = width of tire footprint = 20 in.

If $h > h_{int}$ (Eq 4-11)

$$L_2 = \frac{t_w + 72\text{ in.} + \text{LLDF}(h)}{2}$$

where:

h_{int} = depth at which load from wheels interacts

$$h_{int} = \frac{72\text{ in.} - t_w}{\text{LLDF}} \qquad \text{(Eq 4-12)}$$

Calculation notes:

1. Equations as shown are for h in inches.
2. AASHTO also specifies a surface lane load of 640 lb/ft over a 10-ft lane width. This lane load is ignored in these calculations as it has only a small effect on the total live load and may be added by the engineer if deemed appropriate.
3. The above calculation method assumes that the live load extends over the full diameter of the pipe. This may be conservative for a large-diameter pipe under low fills. To account for this, the calculated live load pressure on the pipe may be reduced by the ratio L_1/OD if the truck is moving across the pipe and L_1 < OD or by L_2/OD if the truck is moving parallel to the pipe and L_2 < OD. OD is the outside diameter of the pipe in inches.

4. For depths of fill less than 2 ft, or for live load magnitudes greater than HS-25, it may be necessary to consider the local live load effects at the crown of the pipe. Such an analysis is beyond the scope of this manual.
5. The previous calculation is for single-axle trucks. Design for tandem-axle trucks may use the same procedures; however, the following substitutions for L_1 should be used if both axles load the pipe at the same time.

$$L_1 = \frac{\text{axle spacing} + t_w + \text{LLDF}\,(h)}{2} \qquad \text{(Eq 4-13)}$$

   Tandem-axle wheel loads are usually lighter than HS20 or HS25 trucks; for example, the AASHTO LRFD Design Tandem has a 12,500-lb wheel load.

6. Rigid pavements dramatically reduce live load effects on pipe. The Portland Cement Association developed a calculation method to consider loads transmitted through concrete pavements (*Vertical Pressure on Concrete Culverts Under Wheel Loads on Concrete Pavement Slabs*, Portland Cement Assoc. Publication ST-65, 1951) that is still in use today and is suitable for computing live loads on PVC pipe under rigid pavements. The same method is also presented in the *Concrete Pipe Design Handbook* published by the American Concrete Pipe Association.

Table 4-2 presents computed live loads for AASHTO HS20 and HS25 single-axle trucks based on an LLDF of 1.15 (for granular fills). The loads shown assume that the load extends over the full diameter of the pipe. This assumption will not be true for large-diameter pipes with shallow covers. Loads for this condition may be lower. See calculation note #3 below Eq 4-12 for guidance on appropriate adjustments.

Also included in Table 4-2 are live loads from Cooper's E80 railroad loading (taken from the Concrete Pipe Design Manual).

Figures 4-4 to 4-6 depict graphically the AASHTO truck and Cooper's E80 railroad live loads shown in Table 4-2.

PIPE STIFFNESS

The inherent strength, regarding external load-carrying capacity, of flexible pipe is quantified as pipe stiffness. Pipe stiffness is measured, according to ASTM D 2412, Standard Test Method for External Loading Properties of Plastic Pipe by Parallel-Plate Loading, at a uniform datum of 5 percent deflection.

Pipe stiffness is defined as follows:

$$PS = F/(\Delta Y) \geq \frac{EI}{0.149r^3} = \frac{6.7EI}{r^3} = \frac{6.71Et^3}{12r^3} = 0.559E\left(\frac{t}{r}\right)^3 \qquad \text{(Eq 4-14)}$$

where:

PS = pipe stiffness, lbf/in./in.

F = force, lb/lin in.

ΔY = vertical deflection, in.

E = modulus of elasticity, psi

I = moment of inertia of the wall cross-section per unit length of pipe $\left(\frac{t^3}{12} \times 1\right)$, in.^{4}/lin in. = in.^{3}

r = mean radius of pipe, in.

t = wall thickness, in.

Table 4-2 AASHTO HS20, HS25, and Cooper E80 live loads (psi)

AASHTO Live Loads		HS20		HS25		Cooper E80 Live Loads			
Depth		W_L		W_L		Depth		W_L	
ft	*m*	*psi*	*kPa*	*psi*	*kPa*	*ft*	*m*	*psi*	*kPa*
2	0.6	13.4	92	16.8	116	3	0.9	16	110
2.5	0.8	9.7	67	12.2	84	4	1.2	14.1	97
3	0.9	7.4	51	9.2	63	5	1.5	12.2	84
4	1.2	4.7	32	5.9	41	6	1.8	10.5	72
5	1.5	3.4	23	4.2	29	7	2.1	9	62
6	1.8	2.6	18	3.2	22	8	2.4	7.7	53
8	2.4	1.6	11	2	14	10	3.0	5.7	39
10	3.0	1.1	7.6	1.4	10	12	3.7	4.6	32
12	3.7	0.8	5.5	1.1	7.6	15	4.6	3.4	23
15	4.6	0.6	4.1	0.7	4.8	20	6.1	2.2	15
20	6.1	0.4	2.8	0.5	3.4	25	7.6	1.5	10
28	8.5	0.2	1.4	0.25	1.8	30	9.1	1.1	7.6
40	12.2	0.1	0.7	0.1	0.7	40	12.2	0.6	4.1

Note: Cooper E80 as defined by AREMA (American Railway Engineering and Maintenance-of-Way Association) in its Manual of Railway Engineering.

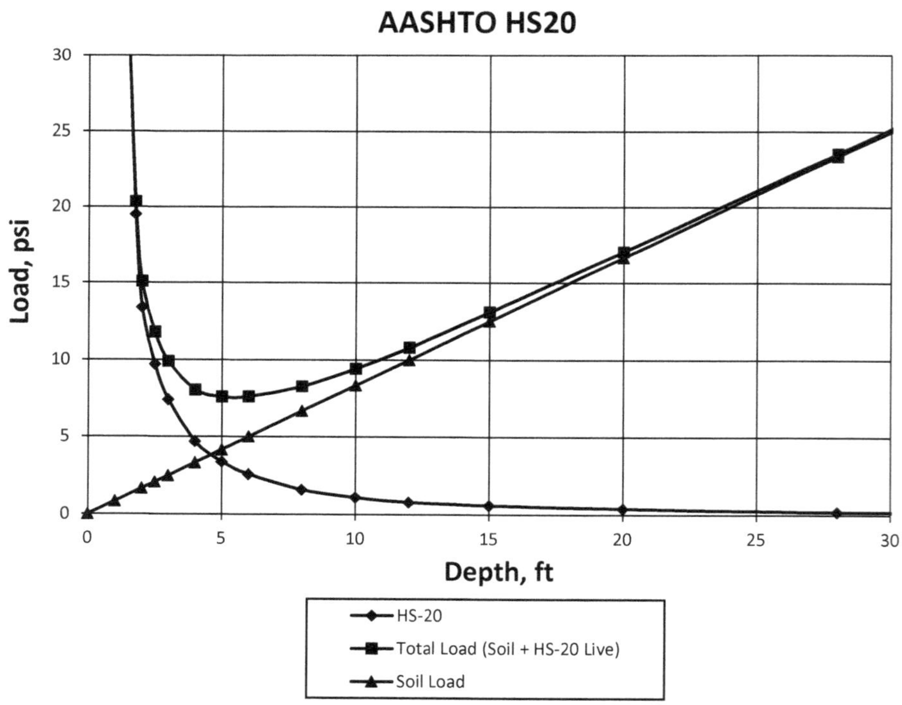

Source: US Bureau of Reclamation

Figure 4-4 AASHTO HS20 live load, soil load (120 pcf), and total load graph

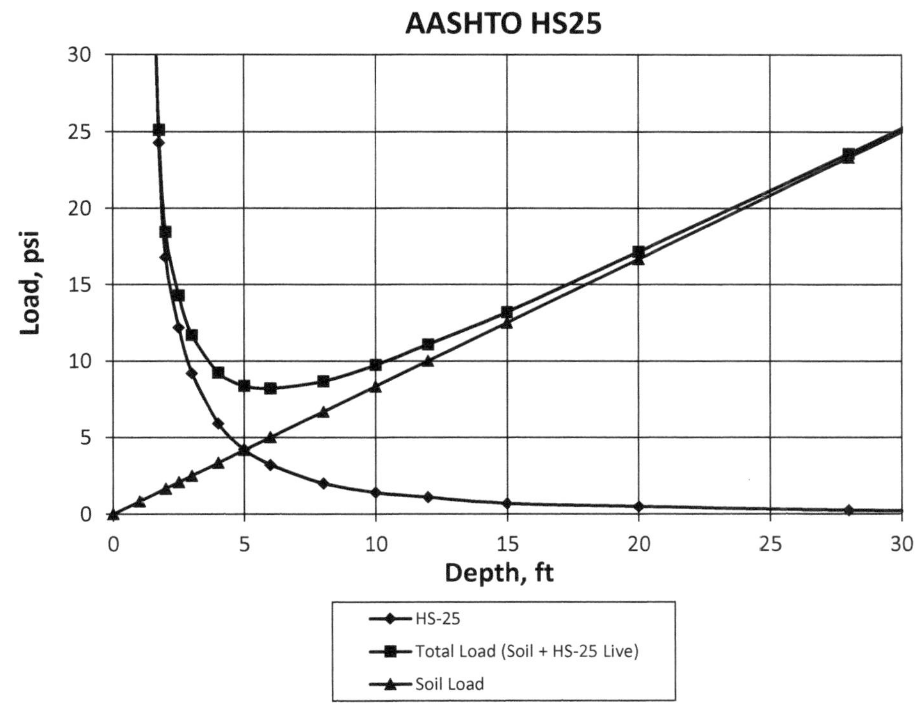

Source: US Bureau of Reclamation

Figure 4-5 AASHTO HS25 live load, soil load (120 pcf), and total load graph

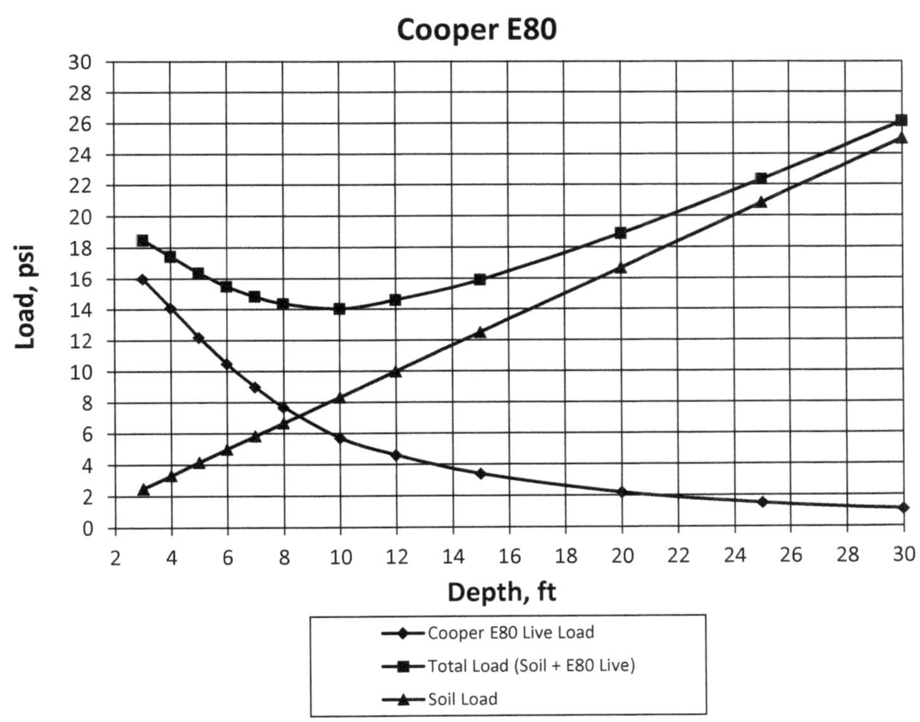

Source: US Bureau of Reclamation

Figure 4-6 Cooper E80 live load, soil load (120 pcf), and total load graph

For PVC pipe with outside diameter controlled dimensions, Eq 4-14 can be simplified further:

$$PS = 4.47 \frac{E}{(DR - 1)^3} \quad \text{(Eq 4-15)}$$

where:

DR = dimension ratio, $\frac{D_o}{t}$
D_o = outside diameter, in.

The resulting PS values for various dimension ratios of ANSI/AWWA C900 PVC and pressure classes for ANSI/AWWA C909 PVCO pipe are shown in Table 4-3.

For pipe with outside diameter controlled dimensions (i.e., DR classifications), Eq 4-2 can be simplified further:

$$\%\frac{\Delta Y}{D} = \frac{k(T_L W_E + W_L + W_S)}{\frac{2E}{3(DR - 1)^3} + 0.061E'} \quad \text{(Eq 4-16)}$$

Time Lag

Buried flexible pipe continues to deflect after the pipe has been initially backfilled. The time lag factor, T_L, relates the immediate deflection of the pipe to the deflection of the pipe after many years. The primary cause of increasing pipe deflection with time is the increase in overburden load. The full load is not realized until 3–12 months after completion of backfilling (or longer). A secondary cause of increasing pipe deflection is the time-related consolidation of the embedment. The soil consolidation is generally of much less significance than the increasing load.

When the operating internal pipe pressure equals or exceeds the external load, the pipe will tend to re-round. For pipes that are pressurized within 3 months of installation, a time-lag factor of 1.0 can be used. Otherwise, a time-lag factor of 1.5 should be used to estimate the deflection that may occur while the pipe is not pressurized. If the pipe will not be pressurized for several years, a time-lag factor of 2.0 should be used to estimate the deflection until pressurized.

Table 4-3 PVC and PVCO pipe stiffness

PVC Stiffness Min. E = 400,000 psi		PVCO Stiffness Min. E = 500,000 psi	
DR	*lbf/in./in.*	*Pressure Class*	*lbf/in./in.*
51	14	PR160, IPS	24
41	28	PR200, IPS	48
32.5	57	PR250, IPS	98
27.5	96	PR150, CIOD	21
26	115	PR200, CIOD	48
25	129	PR250, CIOD	98
21	224	PC165, CIOD	28
18	364	PC235, CIOD	81
14	815	PC305, CIOD	178

Bedding Coefficient

The bedding coefficient, k, reflects the degree of support provided by the soil at the bottom of the pipe. For direct burial conditions, a k value of 0.1 is used.

Modulus of Soil Reaction, *E′*

The pipe deflection depends on the stiffness of the embedment soil at the sides of the pipe. The stiffness is characterized by the modulus of soil reaction, E'. The value of E' depends on the soil classification and percent compaction of the embedment soil. As presented in ASTM D 2774, the embedment soils can be categorized into five classes as shown in Table 4-1. The modulus of soil reaction, E', for the various soil classes are included in Table 4-4. The E' value for each of the soil classes increases with the degree of compaction.

For the E' of crushed rock, Class I, the distinction in Table 4-4 is between uncompacted and compacted. Uncompacted means that the material is loosely placed (dumped) around the pipe with no compaction. Compaction means that some type of compaction equipment is used to densify the soil. The percent compaction is not determined because of the inherent difficulty in obtaining in-place and laboratory densities.

Composite Modulus of Soil Reaction, *E′*

The vertical loads on a flexible pipe can decrease the pipe vertical diameter and increase the horizontal diameter. The horizontal movement develops a passive soil resistance that acts to help support the pipe. While the embedment soil is normally the primary source for the passive resistance, the extent of pipe deflection may also be influenced by the trench walls. Very weak (low modulus) native trench wall soils can influence the pipe deflection, and their lower stiffness may need to be combined with the stiffness of the embedment soil, as a composite E', in order to predict pipe deflection.

The composite E' depends on the soil type and the degree of compaction of the embedment material, the native soil stiffness, the pipe diameter, and the trench width. To calculate the composite E', determine the E'_b of the embedment material, the E'_n of the native soil, and combine them using Eq 4-17. Values of S_c are shown in Table 4-5.

$$E' = S_c\, E'_b \qquad \text{(Eq 4-17)}$$

where:

E' = composite modulus of soil reaction, psi (to be used in Eq 4-2 or 4-16)
S_c = soil support combining factor from Table 4-5, dimensionless
E'_b = E' of the embedment from Table 4-4, psi

To use Table 4-5 for S_c values, the following values must be determined:

E'_n = E' of the native soil at pipe springline elevation from Table 4-6, psi
B_d = trench width at pipe springline, in.
D = pipe diameter, in.

The E'_n for the trench walls can be determined from Table 4-6 or from the values in Table 4-4. The latter choice will be conservative because Table 4-4 E' values are for disturbed soils that have been re-compacted. Native soils will typically have higher E' values because of their natural structure and cementation. Other methods, such as unconfined compression tests or cone penetration test (CPT), to determine an E'_n are available.

Table 4-4 *E′* values, psi

Soil Group (USCS)	Uncompacted	Compacted	
		Moderate 85–90% compaction	High ≥95% compaction
Class I Crushed rock	1,000	6,000	
Class II GW GP SW SP	500	2,000	4,000
Class III GC GM SC SM CL ML CL ML	200	1,000	2,500
Class IV CL ML	100	400	1,500
Class V CH MH OH OL Pt	Do not use		

USCS: Unified Soil Classification System

Notes:

1. Soil Classification in accordance with ASTM D2487.
2. Percent compaction is based on ASTM D698.
3. For information on soil classes, see Table 4-1.
4. *E′* values only valid for cover depths of 50 ft or less. The values must be used with a prism load. These values are only valid for estimating initial deflections. A time-lag factor must be used to predict long-term deflections.

Table 4-5 Values for the soil support combining factor, S_c

	B_d/D					
E'_n/E'_b	1.5	2.0	2.5	3	4	5
0.1	0.15	0.30	0.60	0.80	0.90	1.00
0.2	0.30	0.45	0.70	0.85	0.92	1.00
0.4	0.50	0.60	0.80	0.90	0.95	1.00
0.6	0.70	0.80	0.90	0.95	1.00	1.00
0.8	0.85	0.90	0.95	0.98	1.00	1.00
1.0	1.00	1.00	1.00	1.00	1.00	1.00
1.5	1.30	1.15	1.10	1.05	1.00	1.00
2.0	1.50	1.30	1.15	1.10	1.05	1.00
3.0	1.75	1.45	1.30	1.20	1.08	1.00
≥5.0	2.00	1.60	1.40	1.25	1.10	1.00

Note: In-between values of S_c may be determined by straight-line interpolation from adjacent values.

Embankment Condition

A buried pipe may be constructed as part of a new embankment. For the typical sizes of PVC and PVCO pressure pipe, the most common construction technique is to build the embankment and then excavate a trench for the PVC or PVCO pipe. The previous discussion on trench installation would then apply to the embankment condition. The trench wall stiffness, E'_n, would be the stiffness of the compacted embankment from Table 4-4.

Table 4-6 E'_n based on standard penetration test (SPT) values, psi

Soil description and classification – USCS*	N Value from SPT[†] test (number of blows/ft)					
	0 – 5	5	10	20	30	>50
Clays and silts with less than 30% Sand/gravel CL ML		500	750	1,250	1,500	2,500
Sandy silts, clays With more than 30% sand CL ML Silty or Clayey sand SM SC	Zero	700	1,000	1,500	2,000	3,000
Normally consolidated sands SP, SP-SM, SP-SC		1,000	1,500	2,500	3,000	5,000
Over-consolidated sands SP, SP-SM, SP-SC		2,000	3,000	4,000	5,000	8,000
Gravels, soils with gravel		Typically higher than sands but SPT test very unreliable, use another method				

USCS: Unified Soil Classification System
* Soil Classification based on ASTM D2487 or ASTM D2488
[†] SPT per ASTM D1586.
[‡] For gravel and soils with gravel, estimate E'_n from Table 4-4.

Flowable Fill

Flowable fill is a fluid mixture of Portland cement, soil, and water that hardens into a solid mass. As detailed in Chapter 7, ASTM has several standards relating to the mixing, placing, and testing of flowable fill. ASTM refers to flowable fill as controlled low strength material (CLSM). Flowable fill is typically about two to five times stiffer than compacted soil and thus can provide good support for buried pipe. Flowable fill can be made from a range of materials, including materials obtained from a concrete batch plant to mixtures using native soils excavated from the trench or borrow source. Unless it is a high early-strength mixture, flowable fill should not be backfilled until the day after placement. Flowable fill gains strength after placement so the stiffness for estimating deflection will depend on the mix design of the flowable fill and period of time elapsed before backfill placement. Fresh flowable fill should have a spread of 8–12 in. and the hardened flowable fill should have a compressive strength of 50–100 psi.

The E' for flowable fill depends on the amount of cementitious material, the aggregate, and the time after placement. For estimating deflections using the typical mixture properties, the E' for flowable fill would be about 1,000 psi after one day and 3,000 psi after three days, depending on the mix design. Mix design studies are recommended to determine the properties of selected flowable fill mixtures.

Flotation

Flotation is generally not a design concern for buried PVC and PVCO pipe when it is full of water or where the groundwater table is below the pipe invert. Flotation can occur if groundwater surrounds the pipe and the upward buoyant force is greater than the weight of the pipe and its contents combined with soil prism load above the pipe. Sufficient backfill placed over the pipe will prevent flotation and movement. The recommended depth of cover to prevent flotation over PVC and PVCO pipe is 1.5 times the diameter of the pipe.

EXAMPLES

Design Example No. 1

What is the estimated deflection of a 12-in. diameter DR 18 PVC pipe installed under 10 ft of cover? The native soil is a silty sand (SM). Also assume the following:

Trench width at pipe springline (B_d) = 24 in.
Bedding coefficient (k) = 0.10
Final backfill soil density (w) = 100 lb/ft^3
Time lag factor (T_L) = 1.0
Modulus of elasticity (E') of PVC material = 400,000 psi
No internal operating pressure
No live load or surcharge load

The native SM soil can be used as uncompacted embedment with a E' of 200 psi. Under these conditions, this would be a *basic installation* and deflection calculation is not necessary.

To illustrate the use of a basic installation, the deflection would be estimated as

$$W_E = \frac{wH}{144} = \frac{100 \times 10}{144} = 6.9$$

$$\% \frac{\Delta Y}{D} = \frac{k(T_L W_E + W_L + W_S)(100)}{\left(2E/3(DR-1)^3\right) + 0.061E'}$$

$$\% \frac{\Delta Y}{D} = \frac{0.10(1.0 \times 6.9 + 0 + 0)(100)}{\left(2 \times 400{,}000/3(18-1)^3\right) + 0.061(200)}$$

$$\% \frac{\Delta Y}{D} = \frac{69}{54.3 + 12.2} = 1.0\%$$

The estimated deflection of 1.0 percent is significantly lower than the allowable 5 percent.

Design Example No. 2

What will be the estimated deflection of a 36-in. diameter DR 51 PVC pipe installed under 20 ft of cover but not pressurized for 18 months? The native soil is lean clay (CL) with about 15 percent sand and an average SPT blow count of 10. The embedment soil is an uncompacted pea gravel (GP; Class II). Also assume the following:

Trench width at pipe springline (B_d) = 54 in.
Bedding coefficient (k) = 0.10
Modulus of soil reaction (E'_n) for native soil = 750 psi (Table 4-6)
Final backfill soil density (w) = 120 lb/ft^3
Time Lag factor (T_L) = 1.5 (see discussion on time lag)
Modulus of elasticity (E') of PVC material = 400,000 psi
No internal operating pressure for 18 months
No live load (W_L = 0)
From Table 4-4, E'_b is 500 psi

$$\frac{E'_n}{E'_b} = \frac{750}{500} = 1.5$$

$$\frac{B_d}{D} = \frac{54}{36} = 1.5$$

From Table 4-5, $S_c = 1.30$

$$E' = S_c E'_b = 1.30(500) = 650$$

$$W_E = \frac{wH}{144} = \frac{120 \times 20}{144} = 16.7$$

$$\%\frac{\Delta Y}{D} = \frac{k(T_L W_E + W_L + W_S)(100)}{\left(2E/3(DR-1)^3\right) + 0.061E'}$$

$$\%\frac{\Delta Y}{D} = \frac{0.10(1.5 \times 16.67 + 0 + 0)(100)}{\left(2 \times 400{,}000/3(51-1)^3\right) + 0.061(650)}$$

$$\%\frac{\Delta Y}{D} = \frac{250}{2.1 + 39.65} = 6.0\%$$

The estimated deflection of 6.0 percent exceeds the allowable 5 percent. One alternative would be to increase the pipe stiffness. Recalculate using a DR 25.

$$\%\frac{\Delta Y}{D} = \frac{k(T_L W_E + W_L + W_S)(100)}{\left(2E/3(DR-1)^3\right) + 0.061E'}$$

$$\%\frac{\Delta Y}{D} = \frac{0.10(1.5 \times 16.67 + 0 + 0)(100)}{\left(2 \times 400{,}000/3(25-1)^3\right) + 0.061(650)}$$

$$\%\frac{\Delta Y}{D} = \frac{250}{19.3 + 39.65} = 4.2\%$$

Or recalculate using dumped crushed rock as embedment. Uncompacted crushed rock (Class I) would have a E' of 1,000 psi (Table 4-4).

$$\frac{E'_n}{E'_b} = \frac{750}{1000} = 0.75$$

$$\frac{B_d}{D} = \frac{54}{36} = 1.5$$

From Table 4-5, $S_c = 0.81$ (interpolated)

$$E' = S_c E'_b = 0.81(1000) = 810$$

$$\%\frac{\Delta Y}{D} = \frac{k(T_L W_E + W_L + W_S)(100)}{\left(2E/3(DR-1)^3\right) + 0.061E'}$$

$$\%\frac{\Delta Y}{D} = \frac{0.10(1.5 \times 16.67 + 0 + 0)(100)}{\left(2 \times 400{,}000/3(51-1)^3\right) + 0.061(810)}$$

$$\%\frac{\Delta Y}{D} = \frac{250}{2.1 + 49.4} = 4.8\%$$

Both alternatives result in less than 5 percent estimated deflection. Material pricing would be a factor in the final choice.

REFERENCES

American Association of State Highway and Transportation Officials (AASHTO). 2012 (6th ed.). *AASHTO LRFD Bridge Design Specifications*. Washington, DC: AASHTO.

ASTM International (ASTM). D653. 2014. Standard Terminology Relating to Soil, Rock, and Contained Fluids. West Conshohocken, Pa: ASTM International.

ASTM. 2012. D698. Standard Test Methods for Laboratory Compaction Characteristics of Soil Using Standard Effort (12,400 ft-lbf/ ft^3 (600 kN-m/m^3)). West Conshohocken, Pa: ASTM International.

ASTM D1586. 2018. Standard Test Method for Standard Penetration Test (SPT) and Split-Barrel Sampling of Soils. West Conshohocken, Pa: ASTM International.

ASTM D2166. 2016. Standard Test Method for Unconfined Compressive Strength of Cohesive Soil. West Conshohocken, Pa: ASTM International.

ASTM D2412. 2018. Standard Test Method for External Loading Properties of Plastic Pipe by Parallel-Plate Loading. West Conshohocken, Pa: ASTM International.

ASTM D2487. 2017. Standard Classification of Soils for Engineering Purposes (Unified Soil Classification System). West Conshohocken, Pa: ASTM International.

ASTM D2488. 2017. Standard Practice for Description and Identification of Soils (Visual-Manual Procedure). West Conshohocken, Pa: ASTM International.

ASTM D2774. 2012. Standard Practice for Underground Installation of Thermoplastic Pressure Piping. West Conshohocken, Pa: ASTM International.

ASTM D4253. 2016. Standard Test Methods for Maximum Index Density and Unit Weight of Soils using a Vibratory Table. West Conshohocken, Pa: ASTM International.

ASTM D7382. 2008. Standard Test Methods for Dry Density of Granular Soils using a Vibratory Hammer. West Conshohocken, Pa: ASTM International.

American Water Works Association (AWWA). 2016. C900 Polyvinyl Chloride (PVC) Pressure Pipe and Fabricated Fittings, 4 In. Through 60 In. (100 mm Through 1,500 mm). Denver, Colo.: AWWA.

AWWA. 2016. C909 Molecularly Oriented Polyvinyl Chloride (PVCO) Pressure Pipe, 4 In. (100 mm) and Larger. Denver, Colo.: AWWA.

Howard, A. 2011. Composite E Prime, ASCE Conference Pipelines 2011, San Diego, Calif.

Howard, A. 2015. *Pipeline Installation 2.0*. Relativity Publishing.

PVC Pipe Association. 2012 (5th ed.). *Handbook of PVC Pipe Design and Construction*. Dallas, Tex.: PVC Pipe Association.

Chapter 5

Design for Internal Pressure

This chapter addresses design relative to internal hydrostatic and transient surge pressures.

INTERNAL HYDROSTATIC PRESSURE

Temperature Considerations

The internal pressure capacity of PVC and PVCO pipe is temperature dependent and varies with operating temperature. PVC and PVCO pipe, when manufactured for pressure applications, is rated for sustained, internal pressure, which is often referred to as working pressure or operating pressure. In AWWA's standards, PVC and PVCO pipes are rated for operating pressure capacity at 73.4°F (23°C). As the operating temperature falls below 73.4°F (23°C), the pressure capacity increases to a level higher than the rated pressure class. In practice, this increase provides an additional margin of safety and is usually not otherwise considered in the design. When sustained operating temperatures are above 73.4°F (23°C), the pipe pressure capacity decreases to levels below its pressure class. Table 5-1 shows pipe pressure class de-ratings for sustained elevated operating temperatures. Sustained operating temperatures above 73.4°F (23°C) need to be considered in the proper design of a PVC pressure piping system.

The appropriate temperature coefficient (F_T) found in Table 5-1 should be applied when it is anticipated that the sustained operating temperatures will be above 73.4°F (23°C). A short-term rise in service temperature above 73°F (such as for several weeks during summer) does not require use of temperature coefficients, F_T, for pipeline design. Interpolation can be used for temperatures within the limits of those listed in Table 5-1.

Table 5-1 Thermal de-rating factors for PVC and PVCO pressure pipes and fittings

Sustained Service Temperature		Multiply the Pressure Class at 73.4°F (23°C) by These Factors (F_T)	
°F	(°C)	PVC	PVCO
73.4	(23)	1.00	1.00
80	(27)	0.88	0.87
90	(32)	0.75	0.75
100	(38)	0.62	0.64
110	(43)	0.50	0.53
120	(49)	0.40	0.42
130	(54)	0.30	0.31
140	(60)	0.22	—

Notes:
1. The maximum recommended service temperature for PVC pressure pipe and fittings is 140°F (60°C), while that for PVCO pressure pipe is 130°F (54°C).
2. Elastomeric compounds used in the manufacture of pipe gaskets are generally suitable for use in water with only slight reduction in long-term properties at continuous temperatures listed above.
3. The de-rating factors in Table 5-1 assume sustained elevated service temperatures.

$$WP \leq PC \times F_T \qquad \text{(Eq 5-1)}$$

where:

WP = working pressure, psi
PC = pressure class, psi
F_T = temperature coefficient (Table 5-1)

Adjustments for Time-Dependent Properties

Metal pipes display an insignificant difference between short-term and long-term design strength. Their pressure ratings are based on short-term tensile strength testing. However, for all plastic pipe products, the relationship between stress and strain is significantly influenced by duration of loading and temperature. The capacity to resist failure is reduced as duration of loading is increased for plastics. The time-dependent, viscoelastic nature of plastics requires that the long-term, sustained stress capacity be used instead of short-term yield strength.

In response to an applied stress such as internal pressure, PVC and PVCO pipe will gradually yield to a point and at a rate that depends on the level and duration of applied stress. The rate of creep in response to a constant internal pressure (or any given stress) decreases with the passage of time. Research has shown that PVC pipes held at their pressure ratings continuously for 11.4 years (100,000 hr) will exhibit no perceptible rate of creep at the end of that time. Further tests of PVC pipes under constant pressure, at twice their pressure class, have shown that their quick-burst strength actually increased during the 10-year span as a result of the molecular reorientation that occurred.

The response of PVC and PVCO pipe to applied stress is displayed in Figure 5-1 and is known as the stress regression curve (SR curve). The SR curve is the line drawn through the stress/time failure data points. It does not represent a loss of strength in a PVC pipe with time. The curve, as shown in Figure 5-1, has been plotted with the horizontal axis representing the logarithm of time to permit plotting a greater passage of time—from 10 s to 1.1 million years. If the time axis were plotted on a linear scale (Cartesian coordinates), the variation in hoop stress, which relates directly to applied internal hydrostatic internal

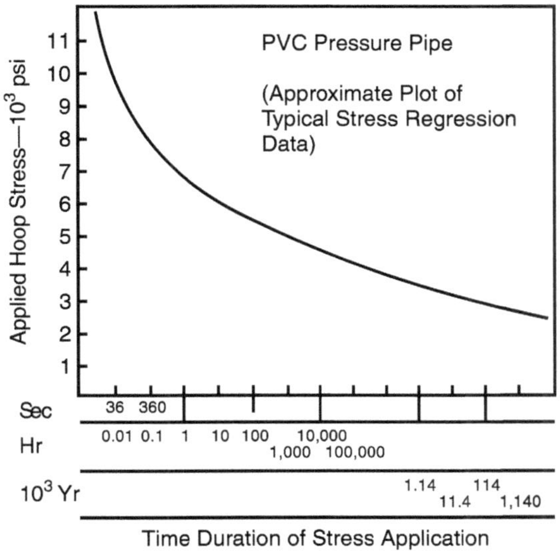

Figure 5-1 Stress regression curve for PVC pressure pipe

pressure, would appear insignificant from 100,000 hr to 500 years. In a practical sense, a long-term response to applied hydrostatic pressure can be based on the hoop stress rating at 100,000 hr in that:

- The response of PVC and PVCO pipe to applied internal hydrostatic pressure or applied hoop stress has essentially stabilized at 100,000 hr when considering the design life of piping systems at 100 years.
- The response of PVC and PVCO pipe to applied hoop stress after 100,000 hr can be accurately determined through testing performed in accordance with ASTM D1598, Standard Method of Test for Time-to-Failure of Plastic Pipe Under Long-Term Hydrostatic Pressure, and through analysis performed as required in ASTM D2837, Standard Method for Obtaining Hydrostatic Design Basis for Thermoplastic Pipe Materials or Pressure Design Basis for Thermoplastic Pipe Products.

The pressure class (PC) for PVC and PVCO pipe, and for all thermoplastic pipe manufactured in North America, is based on the ASTM D2837 standard method and log-log plot of stress regression, commonly termed the stress regression (SR) line. Figure 5-2 displays typical SR lines for PVC and PVCO pressure pipe. The SR line is a plot of the same data plotted on the SR curve; however, the hoop stress is plotted using a logarithmic scale. The log-log plot of long-term stress response data for PVC and PVCO pipe plots as a straight line. Use of the SR line, with constant slope, permits accurate mathematical projection of extremely long-term response to sustained stress. It is possible to project that the estimated life for a PVC pressure pipe, operating at full-rated pressure (i.e., 2,000 psi [13.8 MPa] hoop stress) is between 1×10^{16} or 1×10^{17} hr, or between 1.14×10^{12} and 1.14×10^{13} year. This projected life expectancy also applies to PVCO pipe operating at full-rated pressure (i.e., 3,550 psi [24.5 MPa] hoop stress).

Research published in 1981 by Robert Hucks helped confirm this long-term strength behavior while demonstrating the retention of PVC pipe strength throughout the life period. Figure 5-3 is a log-log plot of the quick-burst hoop stresses as they were recorded by Hucks. All PVC pipe specimens were held at a constant hoop stress of 4,000 psi (27.58 MPa). Actual quick-burst hoop stress values increased from about 8,000 psi (55.16 MPa) to over 10,000 psi (68.95 MPa).

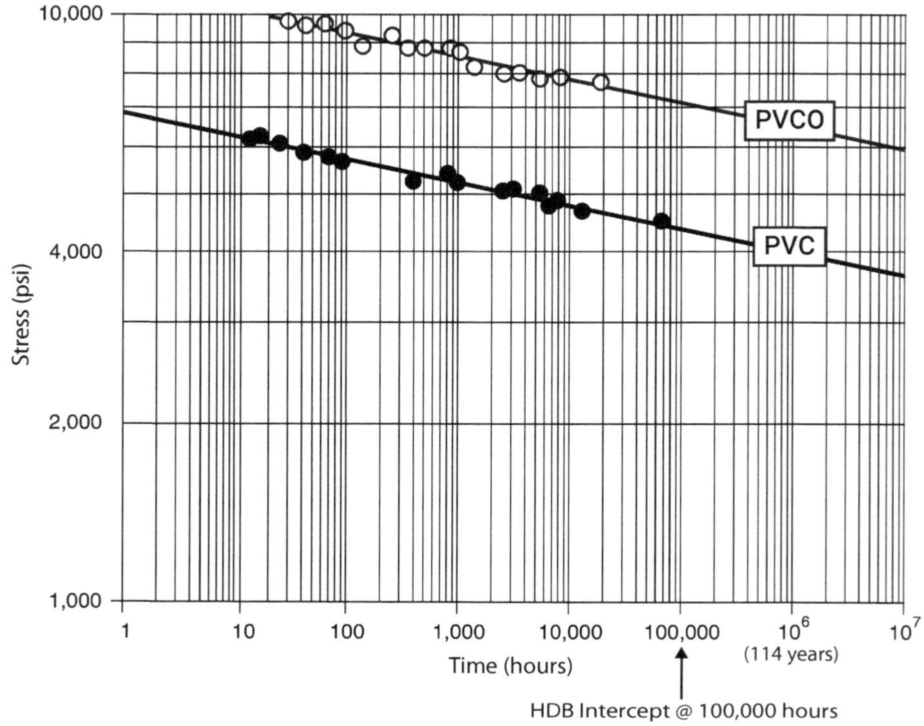

Figure 5-2 Stress regression lines

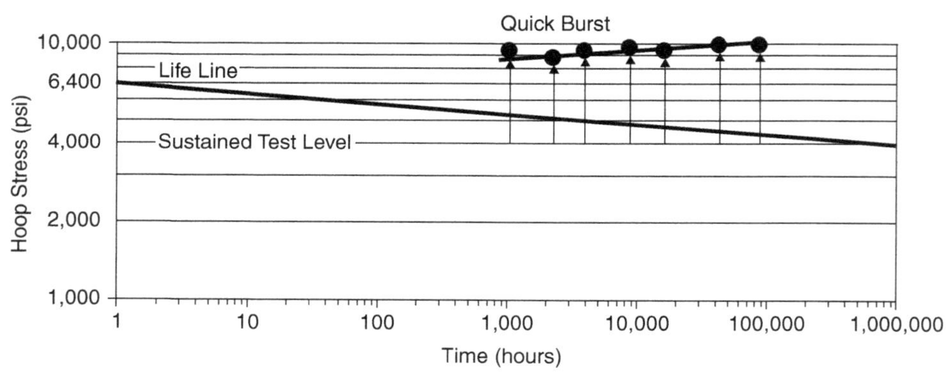

Figure 5-3 Strength and life lines of PVC 12454

All PVC and PVCO pressure pipe manufactured in North America is required to be extruded from PVC compounds for which stress regression lines have been established or for which prequalified ingredients, based on results of industry-wide testing, are used within prescribed limits. More information regarding the requirements and safeguards for PVC and PVCO pressure pipe compounds are provided in Chapter 2.

PVC pipe designed and manufactured for pressure water applications must have a hydrostatic design basis (HDB) of 4,000 psi (27.58 MPa) as defined by ASTM D2837. PVCO pipe designed and manufactured for pressure water applications must have a defined HDB of 7,100 psi (49.0 MPa) as defined by ASTM D2837. Examples using 4,000 psi (27.58 MPa) will be used throughout this manual. However, it should be recognized that other HDBs may be used for specific products such as injection-molded PVC pressure fittings and PVCO pipe.

The sustained operating pressure, rather than occasional surge pressures or recurring transient pressures, usually governs the wall thickness requirements for properly designed and operated PVC and PVCO pressure pipelines. Both PVC and PVCO pipe can withstand high occasional pressure surges and typical transient pressures.

For example, a Pressure Class 235 psi (PC 235) PVC pipe (ANSI/AWWA C900) is designed to withstand a constant internal pressure of 235 psi (1,620 kPa) for in excess of 1,000 years. That same PC 235 psi PVC pipe has a required minimum quick-burst pressure of 755 psi (5.21 MPa) and can withstand over 20 million 30-psi pressure cycles at a sustained operating pressure of 150 psi.

All PVC and PVCO pipe pressure classes (PCs) include a long-term safety factor of at least 2.0. That 2.0 safety factor is applied to the material's ASTM D2837, long-term strength capacity, or HDB.

When qualifying a joint design, a manufacturer has two options. The first option is to thicken the bell to maintain the dimension ratio (DR) of the pipe. The second option is to test to verify that joint assemblies qualify for an HDB category of 4,000 psi (27.58 MPa). The requirements for joint design qualification are published in ASTM D3139, Standard Specification for Joints for Plastic Pressure Pipes Using Flexible Elastomeric Seals.

Pressure Classifications

In the definition of PVC and PVCO pipe's pressure class, the hydrostatic design stress (HDS or *S*) rather than the HDB and safety factor (SF) can be used in the calculations. Figure 5-4 provides an illustration of internal pressure, design stress, and pipe wall thickness. The design stress is simply the value obtained when HDB is divided by the SF.

$$S = \frac{\text{HDB}}{\text{SF}} \qquad \text{(Eq 5-2)}$$

where:

S = design hoop stress, psi
HDB = hydrostatic design basis, psi
SF = safety factor

The design stress (S) is the allowable sustained hoop stress used in the calculation of pipe pressure class. PVC and PVCO pipe pressure class is calculated in accordance with standard practice defined by the International Standards Organization (ISO) Equation R161-1960 based on the work first published by Lamé in 1852:

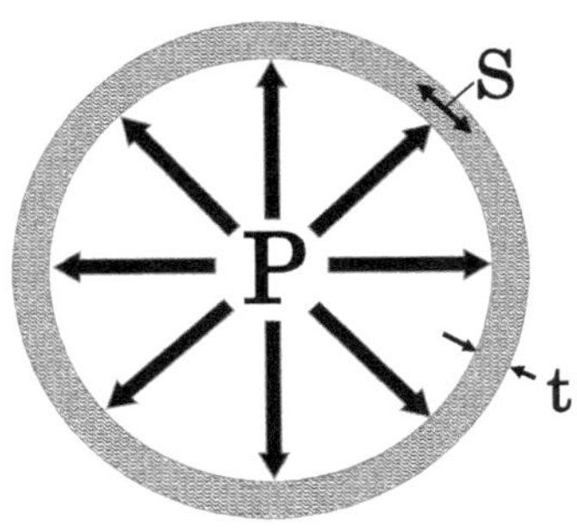

Figure 5-4 Diagram showing internal pressure, design stress, and pipe wall thickness

$$\frac{2S}{P} = \frac{D_o}{t} - 1 = \text{DR} - 1 \qquad \text{(Eq 5-3)}$$

where:

S = design hoop stress, psi
P = pressure rating, psi
D_o = average outside diameter, in.
t = minimum wall thickness, in.
DR = D_o/t, dimension ratio

The ISO Equation can be transposed to a form whereby it can be better utilized in defining pressure capacities of PVC pipe:

$$P = \frac{2S}{\text{DR} - 1} \qquad \text{(Eq 5-4)}$$

Equation 5-4 can be used to determine what internal pressure would create a given hoop stress. Or conversely, the equation can determine the magnitude of hoop stress generated by a given internal pressure.

ANSI/AWWA C900 classifies pipe in accordance with its pressure class (PC). The ISO equation is used to calculate the PC:

$$\text{PC} = \frac{2 \times \text{HDB}}{(\text{DR} - 1) \times \text{SF}} \qquad \text{(Eq 5-5)}$$

where:

PC = pressure class, psi
DR = dimension ratio, D_o/t
D_o = average outside diameter
t = minimum pipe wall thickness
HDB = hydrostatic design basis, 4,000 psi for PVC and 7,100 psi for PVCO (ASTM D2837)
SF = safety factor, 2.0, provides for a high degree of safety and accounts for installation and operation variables, including limited surge effects

PC values for ANSI/AWWA C900 pipe are listed in Table 5-2 together with their minimum burst-test pressures. The PC values are the recommended safe sustained working pressure (WP) limits for ANSI/AWWA C900, PVC pipe. They assume that the pipe system's sustained operating temperature is at or below 73.4°F (23°C).

The ANSI/AWWA burst-test pressures listed in Table 5-2 are conservative measures of the pipe's minimum short-duration pressure limit for temperatures at or below 73°F (23°C). The burst-test is a manufacturing quality control test, and the required pressures should never be used for testing installed pipe.

For PVCO pipe, pressure classes are also calculated using Eq 5-4. The only differences lie in the value for HDB. PVCO has an HDB = 7,100 psi, and the term D_o/t is substituted for DR.

INTERNAL SURGE PRESSURE

Transient pressure surges are a result of any significant deviation from the normal steady-state operation of a piping system and are not sustained events. Transient pressure surge duration depends on the length and configuration of the piping system.

Table 5-2 Pressure classes and required minimum burst-test pressures for PVC pipe (ANSI/AWWA C900)

	Pressure Class		Burst-Test Pressure	
DR	*psi*	(*kPa*)	*psi*	(*kPa*)
51	80	(550)	260	(1,800)
41	100	(690)	320	(2,180)
32.5	125	(860)	410	(2,760)
27.5	150	(1,030)	485	(3,280)
26	160	(1,100)	515	(3,450)
25	165	(1,140)	535	(3,690)
21	200	(1,380)	640	(4,350)
18	235	(1,620)	755	(5,210)
17	250	(1,720)	800	(5,500)
14	305	(2,100)	985	(6,800)

Note: Burst pressures listed should not be used for field testing.

ASTM D1599, Standard Test Method for Resistance to Short-Time Hydraulic Pressure of Plastic Pipe, Tubing, and Fittings, is used to measure a pipe's short-time internal pressure capacity. The required minimum failure time is 60 s and provides a conservative measure of the pipe's ability to handle transient pressure surges that typically are of shorter duration. The ASTM D1599 burst test requirements for PVC and PVCO pipe are given in their respective ANSI/AWWA standards, C900 and C909. Table 5-2 includes the required minimum burst pressures for PVC and PVCO pipe.

The required minimum burst pressures for PVC and PVCO pipe can be approximated by multiplying the PC by 3.2, i.e., PVC and PVCO pipes' surge pressure capacities are $\geq 3.2 \times$ PC.

When a flowing liquid in the pipe changes velocity, a surge pressure or water hammer is generated. The kinetic energy of the moving fluid must be converted to potential (stored) energy and ultimately dissipated through frictional losses in the fluid and in the pipe wall.

Some of the more common causes of significant pressure surges are as follows:

- the rapid opening and closing of valves;
- uncontrolled starting and stopping of pumps;
- rapid changes in turbine speeds;
- liquid column separation;
- rapid movements of entrapped air, and
- improperly sized pressure-reducing valves.

For proper PVC and PVCO pipe design, surges should be divided into two categories: occasional surges and recurring (cyclic) surges. Surge pressures that occur during periodic maintenance operations or as a result of emergencies, such as power failures, sudden valve closures, or system component failures/malfunctions, should be treated as occasional surges. The allowable occasional surge pressure should not exceed the pipe's minimum burst pressure (Table 5-2) divided by a factor of safety.

The magnitude of occasional and cyclic transient surges can be calculated using the elastic wave theory of surge analysis. In addition, the geometry and boundary conditions of a pipe system can be complicated, meriting the use of refined techniques like those developed by Streeter and Wylie. Computer software can assist with the hydraulic and transient analysis of complex piping systems.

The surge pressure is related to the change of the flow velocity, while the rate of travel of the pressure wave is related to the speed of sound in the fluid (modified by the piping material).

The wave velocity is given by the following equation:

$$\pm = \frac{4{,}660}{\sqrt{1 + (kD_i / Et)}} \qquad \text{(Eq 5-6)}$$

where:

α = wave velocity, ft/s
k = fluid bulk modulus (300,000 psi for water)
D_i = pipe inside diameter, in.
E = modulus of elasticity of the pipe material (400,000 psi for PVC, 500,000 psi for PVCO)
t = wall thickness, in.

Substituting dimension ratio DR for ($D_i/t + 2$)

[Note that DR = $D_o/t = (D_i + 2t)/t = D_i/t + 2$]

$$\pm = \frac{4{,}660}{\sqrt{1 + (k(\text{DR} - 2) / E)}} \qquad \text{(Eq 5-7)}$$

The maximum pressure surge may be calculated using

$$P = \frac{\pm V}{(2.31)g} \qquad \text{(Eq 5-8)}$$

where:

P = pressure surge, psi
V = rapid or instantaneous velocity change, ft/s^2
g = acceleration due to gravity, 32.2 ft/s^2

Example: A flow of 2 ft/s is suddenly stopped in a 16-in. Pressure Class 235 (DR 18) PVC pipe.

Calculate the expected maximum surge pressure.

Solution: First calculate the wave velocity:

Substitute the given values into Eq 5-7
k = 300,000 psi
DR = 18
E = 400,000 psi

$$\pm = \frac{4{,}660}{\sqrt{1 + (300{,}000(18 - 2) / 400{,}000)}}$$

Solving yields α = 1,292 ft/s.

Now substitute α and V into Eq 5-8.

Solving for P yields $P = \dfrac{(1{,}292)(2)}{(2.31)(32.2)} = 35\ \text{psi}$

Pressure surges in PVC pipe of different dimension ratios in response to a 1-ft/s (0.3-m/s) instantaneous flow velocity change are listed in Table 5-3.

For PVCO pipe, the surge pressure for an instantaneous 1-ft/s flow velocity change can be approximated by multiplying the previous PVC pipe pressure surge associated with the same PC by 0.81.

Table 5-3 PVC pipe pressure surge for each 1-ft/s (0.3-m/s) instantaneous flow velocity change

	Pressure Class (PC)		Pressure Surge Per 1 ft/s	
DR	*psi*	*(kPa)*	*psi*	*(kPa)*
51	80	(550)	10.8	(74)
41	100	(690)	11.4	(79)
32.5	125	(860)	12.8	(88)
27.5	150	(1,030)	14.0	(96)
26	160	(1,100)	14.4	(99)
25	165	(1,140)	14.7	(101)
21	200	(1,380)	16.0	(110)
18	235	(1,620)	17.4	(120)
17	250	(1,720)	17.9	(123)
14	305	(2,100)	19.8	(137)

Design for Occasional Pressure Surges

Surge pressures that occur during periodic maintenance operations or as a result of emergencies, such as power failures, sudden valve closures, or system component failures/malfunctions, are treated as occasional surge pressures (P_{OS}).

The sum total of the design working pressure (WP) and maximum anticipated occasional surge pressure (P_{OS}) should be less than or equal to the pipe's minimum burst pressure, as required in the pipe product standard. To be consistent, the same safety factor of 2.0 used to reduce the sustained stress capacity (HDB) to a sustained design stress (HDS) is generally applied to the short-term burst pressure, thus reducing it from 3.2 × PC to 1.6 × PC. The resulting design equation for handling occasional surge pressures is Eq 5-9.

$$\mathrm{WP} + \mathrm{P_{OS}} \leq 1.60\,\mathrm{PC} \times F_T \qquad \text{(Eq 5-9)}$$

where:

- WP = working pressure, psi
- P_{OS} = occasional surge pressure, psi
- PC = pressure class, psi
- F_T = temperature coefficient (Table 5-1)

The safe occasional pressure surges (P_{OS}) and corresponding permissible rapid velocity changes are listed in Table 5-4.

For PVCO pipe of the same PC, the allowable ΔVs can be approximated by multiplying the values shown in Table 5-4 by 1.25.

Design for Recurring (Cyclic) Surges

Recurring (cyclic) surge pressures (P_{RS}) are those that occur several times daily or several times hourly. They are most often associated with pumps that automatically cycle on and off, and/or valves that operate automatically. Piping materials, including PVC, may eventually fatigue if exposed to continuous cyclic surging of a sufficiently high frequency and stress amplitude.

Table 5-4 Safe maximum occasional surge pressures and allowable sudden changes in water velocity for ANSI/AWWA PVC pipe operating at working pressures (WP) expressed as percent of nominal pressure class (PC)*

	Pressure Class (PC)	Occasional Surge Pressure Capacity (1.6 x PC)	Corresponding Sudden ΔV with WP = % of PC			
			100%	80%	60%	40%
DR	*psi (kPa)*	*psi (kPa)*	*fps (m/s)*	*fps (m/s)*	*fps (m/s)*	*fps (m/s)*
51	80 (550)	128 (880)	4.7 (1.4)	6.3 (1.9)	7.9 (2.4)	9.5 (2.9)
41	100 (690)	160 (1,100)	5.3 (1.6)	7.0 (2.1)	8.8 (2.7)	10.6 (3.2)
32.5	125 (860)	200 (1,380)	5.9 (1.8)	7.8 (2.4)	9.8 (3.0)	11.8 (3.6)
27.5	150 (1,030)	240 (1,655)	6.5 (2.0)	8.6 (2.6)	10.8 (3.3)	12.9 (3.9)
26	160 (1,100)	256 (1,765)	6.5 (2.0)	8.7 (2.7)	10.9 (3.32)	13.1 (4.0)
25	165 (1,140)	264 (1,820)	6.7 (2.0)	9.0 (2.7)	11.2 (3.41)	13.5 (4.11)
21	200 (1,380)	320 (2,210)	7.5 (2.3)	10.0 (3.05)	12.5 (3.81)	15.0 (4.57)
18	235 (1,620)	376 (2,590)	8.1 (2.5)	10.8 (3.29)	13.5 (4.11)	16.2 (4.94)
17	250 (1,720)	400 (2,760)	8.4 (2.6)	11.2 (3.41)	14.0 (4.27)	16.8 (5.12)
14	305 (2,100)	488 (3,365)	9.2 (2.8)	12.3 (3.75)	15.4 (4.70)	18.5 (5.64)

Source: ANSI/AWWA C900-16

* The surge pressure tolerances stated in this table apply to pipe operating at or below 73°F and not to non-PVC system components, which may have lesser tolerances. The design should consider possible system reactions and their potential effect on all system components.

A pipe may sometimes be subjected to net negative internal pressure because of the individual or combined effect of internal negative transients and external forces (such as water table). When this situation exists, refer to the supplier for information on the hydraulic collapse resistance of the pipeline.

Water distribution mains generally do not pose a fatigue-failure risk due to their smaller surge pressure amplitudes. In an independent survey of 228 water utilities sponsored by the American Water Works Association Water Research Foundation, 80 percent reported pressure fluctuations (surges) below 20 psi, with less than five percent indicating they had pressure fluctuations greater than 40 psi.* Most water transmission lines are designed to experience no more than a few pump cycles per day. Lower surge pressures and/or less frequent surges reduce the risk of pipe failure from cyclic surge pressure fatigue.

As detailed in the previous section, occasional surge pressures can include rapid or instantaneous velocity changes. Proper design, however, precludes rapid velocity changes for recurring surges. Recurring surges are minimized through the use of suitable protection devices or operating procedures. All new pipelines and pipe systems should be equipped with devices to prevent rapid movements and to minimize pressure surges.

Designing a pipe system to handle cyclic loads due to pressure surges or cyclic pressure fluctuations can be complex. Large pressure surges should not occur, and pressure cycles should not be frequent. Using any of the surge pressure control techniques listed in the following section or other surge control techniques substantially decreases the amplitude of recurring pressure surges and thereby lowers the overall cost of the pipeline or pipe system by reducing the wall thicknesses required for the pipe, fittings, and other appurtenances. Consequently, the amplitude of recurring surge pressures is not the result of instantaneous flow velocity changes.

Pipe designers may also use computer software to analyze the locations and extent of recurring pressure surges. When using computer software, the designer should ensure that he or she understands the basis for the inputs and outputs of the program. The designer can use this information to select the most cost-effective PVC or PVCO pipe PC for each section of the pipeline.

* Moser, A.P. and Kellogg, G. 1994. *Evaluation of Polyvinyl Chloride (PVC) Pipe Performance.*

Empirical correlations for estimating PVC pipe fatigue failure response to cyclic loading have been derived from cyclic testing of pipe specimens. S-N curves (stress range vs number of cycles to failure) for ANSI/AWWA C900 PVC pipe have been generated and are familiar to most engineers. Three variables—mean (average) stress, stress amplitude, and number of cycles to failure—are considered in order to properly design a PVC pipe to handle recurring (cyclic) surge pressures.

Equations 5-10 and 5-11 were developed at Utah State University to predict the fatigue failure of PVC pressure pipe. These equations provided the best fit with the fatigue failure test results.

$$\sigma_{amp} = \sigma_{mean}\left(0.01637 \cdot \log(NCF) - 0.1697\right) + \frac{15362}{\log(NCF)} - 1782 \qquad \text{(Eq 5-10)}$$

or

$$\sigma_{mean} = 10{,}000{,}000\frac{\left(500 \cdot \sigma_{amp} \cdot \log(NCF) - 7681285 + 891471 \cdot \log(NCF)\right)}{\log(NCF) \cdot \left(81{,}828{,}176 \cdot \log(NCF) - 848{,}462{,}705\right)} \qquad \text{(Eq 5-11)}$$

where:

σ_{mean} = mean stress, psi
σ_{amp} = stress amplitude (zero to peak), psi
NCF = number of cycles to failure

The plot of this equation for various stress amplitudes is presented in Figure 5-5. It shows the positive/negative line and the 2,000 psi peak stress line. Typical designs should be made between these two lines. This design chart is used in the design of PVC sewage force mains and PVC turf irrigation systems if frequent surges of large magnitude are not controlled. Figure 5-5 is also useful for checking a PVC pipe's ability to withstand the many smaller magnitude surges that are typical in a municipal water system.

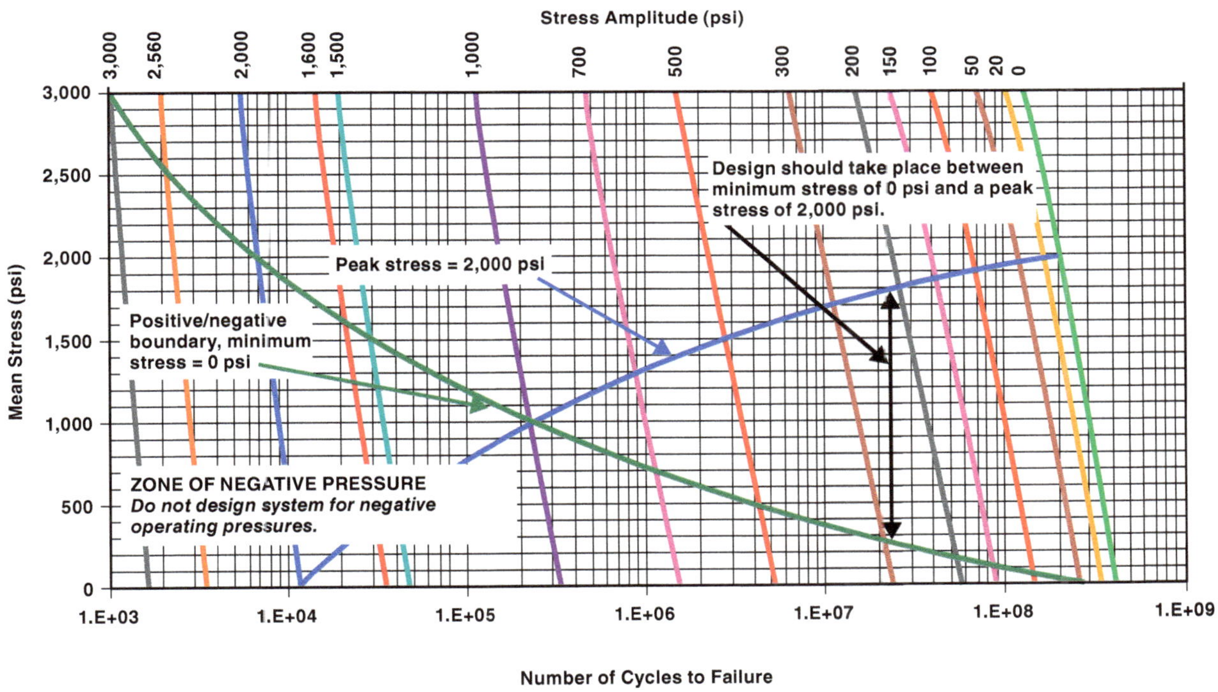

Source: Adapted from Uni-Bell PVC Pipe Association

Figure 5-5 Recurring (cyclic) design graph for PVC pressure pipe

Common Surge Pressure Control Techniques

The designer is advised to keep in mind that the pipe is part of a larger system and while PVC or PVCO pipe may be able to handle high surge pressures, surges could adversely affect other system components.

Surge control concepts are familiar to those who design, install, test, and operate piping systems. A variety of surge control techniques are commonly used to protect the system and keep surge pressures in check. They include but are not limited to the following:

Surge tanks or air chambers. A closed unit containing air and water, sometimes separated by a diaphragm or bladder. The air is under pressure, allowing control of both positive and negative surges in high-pressure systems by allowing flow into and out of the unit.

Variable speed drives (VSDs). VSDs can be used to reduce fatigue damage from surge events. The VSD allows the pump speed to increase slowly to achieve slow line filling, which enables air removal without damaging the pipeline. The VSD also provides flexibility of operation for a process where flows can be increased for future needs without changing the equipment.

Soft starters. Soft starters' primary purpose is to reduce the electrical load on the power supply to a facility. They are widely deployed in many pumping stations for this purpose, particularly where the pump station is at the end of a long power transmission line. They have some but not all features of a VSD. They control ramp speed up and down to some extent. Therefore, they do provide benefits in reducing surge in some applications.

Pump and motor flywheels. Pumps and motors with flywheels decelerate slowly and minimize surges generated. Flywheels in pumping applications are sometimes not the best option. A flywheel will increase the starting current of a pump and thus there will be an increase in cost of motor starters, soft starters, or VSDs.

Slow-closing valves. These are adjustable and may be mechanically or hydraulically operated to minimize abrupt variation in flow velocity. Some employ a two-stage closing process whereby the valve is closed to a 15–20 percent open position rapidly and then the last closure is over an extended period. Similarly, the valve opening is a two-stage process.

Non-slam check valves. The type of check valve used has a great bearing on the pressure transients that can occur in a system. The use of nozzle-type check valves is increasing. The use of this type of valve is but a small investment in reducing risk and fatigue damage.

Pressure relief valves. Relief valves come in a variety of designs. Conventional spring-loaded relief valves generally do not operate fast enough to relieve a pressure wave as it passes. Specialized relief valves have been designed for use in the water industry. They are direct acting valves designed to operate in milliseconds. It can be a spring-loaded disc with no guides that require maintenance to ensure that the valve operates.

Standpipes. These are tanks that are open to the atmosphere and function in a manner similar to a surge tank for low pressure.

Surge tanks with one-way outlets. These allow water to enter the line during negative surges and allow no return on positive surges (useful for negative pressures only).

Surge Pressure Design Example

The following design example illustrates the proper use of Figure 5-5.

Given:

PVC pipe: DR 18, PC = 235 psi
Working pressure (WP) = 160 psi
Anticipated cycles per day = 55
Factor of safety = 2
Design life = 100 years
Anticipated recurring surge pressure (P_{RS}) = ±30 psi (determined by designer)*
Anticipated occasional instantaneous change in flow velocity = 7 ft/s
Sustained operating temperature: ≤73°F

Design analysis:

Since sustained operating temperature does not exceed 73°F, F_T = 1 (Table 5-1) and pressure-rating reduction is not necessary, PC = 235 psi.

Check working pressure versus pressure class (PC)

Working pressure (WP) = 160 psi is less than PC of 235 psi.

Therefore, PC 235, PVC pipe is suitable for the WP of 160 psi.

Use Table 5-3 and the occasional surge pressure design velocity of 7 ft/s to calculate the maximum occasional surge pressure (P_{OS}).

$$P_{OS} = 17.4 \text{ psi} / (1 \text{ ft/s}) \times 7 \text{ ft/s} = 122 \text{ psi}$$

Per Eq 5-9, to provide for a safety factor of 2.0, the sum of the design working pressure (WP) and maximum anticipated occasional surge pressure (P_{OS}) must be less than or equal to the pipe's PC × 1.6.

Anticipated occasional maximum surge pressure (WP + P_{OS}):

$$WP + P_{OS} = 160 \text{ psi} + 122 \text{ psi} = 282 \text{ psi}$$

Occasional surge pressure capacity that provides a 2.0 factor of safety (1.60 × PC × F_T):

$$1.60 \times PC \times F_T = 1.60 \times 235 \text{ psi} \times 1 = 376 \text{ psi}$$

Use Eq 5-9 to compare the occasional maximum design surge pressure to the pipe's design pressure capacity.

$$282 \text{ psi} < 376 \text{ psi}$$

Therefore, PC 235, PVC pipe is suitable for the occasional surge pressure of 282 psi.

Use Figure 5-5 to check the pipe's suitability to handle the anticipated recurring/cyclic surge pressures.

In order to use Figure 5-5, the design working pressure (WP) and recurring/cyclic surge pressure (P_{RS}) amplitude must be expressed in terms of their associated hoop stresses using Eq 5-4.

* In pipelines where water or wastewater is being pumped from one point to another (from a source to a reservoir, for example), a cyclic pressure transient analysis should be conducted by an experienced professional engineer, and recurring surges should be minimized through the use of suitable protection devices and/or operating procedures. This example uses a cyclic surge pressure of ±30 psi—a value that is conservative for pipelines that employ any of the cost-effective surge control techniques listed in this manual. Recurring rapid or instantaneous velocity changes are preventable and should not be the basis for estimating recurring/cyclic pressures. Excessive surge pressure estimates will result in over-design and could exacerbate surge issues in the system as well as raise energy costs.

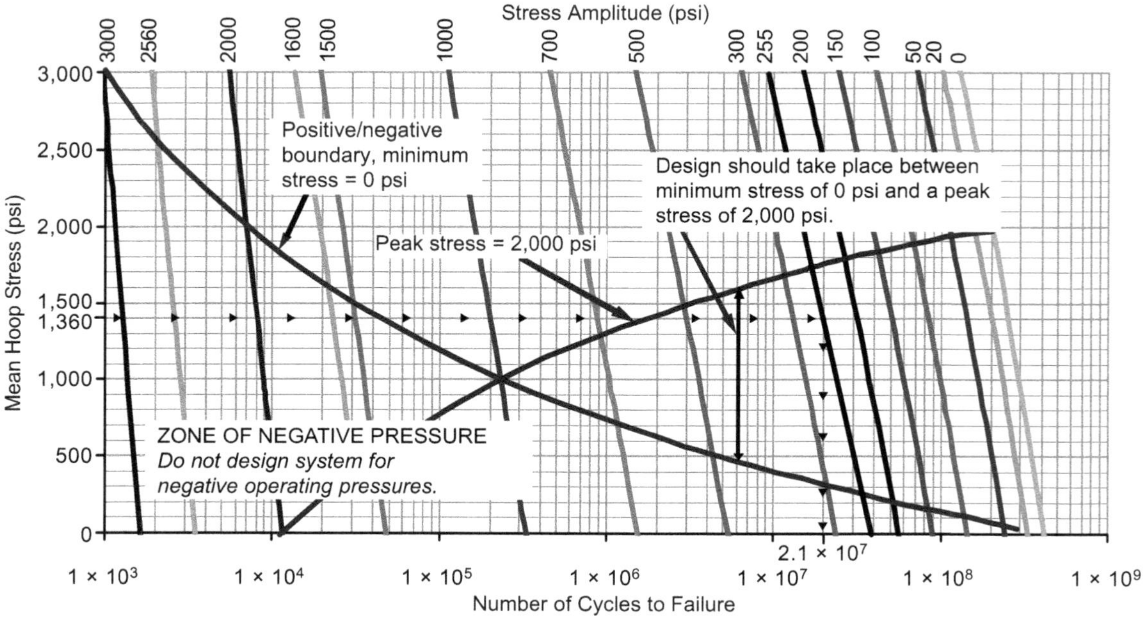

Source: ANSI/AWWA C900-16 (Adapted from Moser, A.P., and G. Kellogg. Evaluation of Polyvinyl Chloride (PVC) Pipe Performance, Denver, Colo.)

Figure 5-6 Recurring (cyclic) design graph example

$$\text{Mean hoop stress} = \frac{\text{WP} \times (\text{DR-1})}{2} = \frac{160 \text{ psi} \times (18 - 1)}{2} = 1{,}360 \text{ psi}$$

$$\text{Recurring surge stress amplitude} = \frac{\text{PRS} \times (\text{DR} - 1)}{2} = \frac{30 \text{ psi} \times (18 - 1)}{2} = 255 \text{ psi}$$

The number of recurring pressure cycles to failure is determined by using Figure 5-5 and the above stress values. This is shown in Figure 5-6.

Number of cycles to failure (from Figure 5-6) = 2.1×10^7 cycles

The projected number of surge pressure cycles to failure should exceed the anticipated number of pressure cycles over the design life of the pipeline.

Anticipated pressure cycles over the design lifetime = 55 cycles/day × 365 days/year × 100 years = 2.0×10^6 cycles

Check the projected versus anticipated number of pressure cycles:

2.1×10^7 cycles > 2.0×10^6 cycles

Therefore, PC 235, PVC pipe is suitable for the recurring/cyclic surge pressures.

In this example, the allowable number of pressure surges exceeds the anticipated number of cycles by a factor of 10.

The PC 235 psi, DR 18, PVC pipe is a conservative selection and provides a safety factor ≥2.0 in all three categories for internal pressure design: static working pressure, occasional transient pressures, and recurring/cyclic pressures.

PVCO pipe has been shown to perform as well as or better than PVC pipe under cyclic pressure conditions. Therefore, Figure 5-5 can also be used as a conservative basis for designing PVCO pipe.

INJECTION-MOLDED PVC FITTINGS

ANSI/AWWA C907 injection-molded PVC pressure fittings use the same principles stated previously for PVC pipe. The pressure capacities of PVC pressure pipe and fittings are based on long-term pressure tests.

Within the limits allowed by the standard, a manufacturer can choose to make injection-molded fittings from one of a number of compounds and employ a variety of machines and mold designs. The total effect of these manufacturing decisions is specific to one size and type of fitting. Because of this, the long-term hydrostatic pressure tests are performed on the fittings themselves rather than the PVC compound. Each size of each configuration of fitting is subjected to regression analysis in the same manner as pipe compounds, except that internal pressure is substituted for hoop stress.

ANSI/AWWA C907 requires that PVC injection-molded fittings have a demonstrated long-term pressure strength equivalent to that of Pressure Class 235 PVC pipe. PC 235, injection-molded PVC fittings may be used with PVC pipe having PCs of 235 psi or less. That is, they can be used with PVC pipes having DRs equal to or greater than DR 18.

If the PVC fittings will be operated at sustained temperatures above 73.4°F (23°C), the pressure class should be reduced accordingly by applying the appropriate factor from Table 5-1.

FABRICATED PVC FITTINGS

PVC pressure fittings may also be fabricated out of sections of PVC pipe. The design of these fittings for internal pressure is identical to that for the pipe. The application and operating conditions should be used to determine the appropriate DR of the fabricated fitting's pipe components. In all cases, PVC fabricated fittings should be made from a pipe having a DR that is less than or equal to the DR requirement for the pipe system. For example, if a designer determines that PC 150 (DR 27.5) pipe is suitable for a transmission main, any fabricated fittings should be made of PC 150 (DR 27.5) PVC pipe or thicker-wall, higher-PC, PVC pipe components. Designers may prefer using thicker-wall, higher-PC, fabricated PVC fittings than that of the adjoining pipe to safely handle the stresses unique to a fitting's geometry.

Specific details on the manufacturing and testing requirements for fabricated PVC fittings are described in ANSI/AWWA C900.

AIR REMOVAL AND VACUUM PREVENTION

Air valves are typically used on sewer force mains and water transmission lines to optimize performance and provide for safety but may not be needed on smaller water distribution pipelines, where hydrants and service lines provide means for venting entrapped air. Air pockets at high points along any pipeline can be detrimental if not controlled. Such air pockets

- reduce cross-sectional flow area, affecting pump efficiency and increasing energy costs;
- magnify surge pressures, which can damage pipe and pipeline components;
- create unsafe conditions when tapping pressurized pipe; and
- will prematurely wear pump equipment and may cause other equipment like control valves to operate erratically.

Because air is 300,000 times more compressible than water, large air pockets can provide enough stored energy to extend post-fracture cracks. In extreme cases, long-running cracks may occur in pipelines with continuous joints (i.e., not segmented) if a very large amount of entrapped air is present. Many operational problems, especially at the time of initial startup, including broken pumps, valves, and pipe, as well as faulty instrumentation readings, result from failure to expel air from the line. Proper air removal will safeguard any pipeline from air-related problems.

Valves for Air Removal

The primary functions of pipeline air valves are to:

1. release air displaced by water as pipelines are filled,
2. admit air to displace the water released when pipelines are drained, and
3. release air that comes out of solution during normal pipeline operation.

The first two functions are accomplished with large orifice "air and vacuum" valves. The third function is accomplished with small orifice "air release" valves. "Combination air" valves incorporate multiple orifices to accomplish all three functions in one apparatus. Combination air valves act as both air release and air/vacuum valves by continuously releasing small volumes of air. Custom combinations of the two valve styles can also be assembled where standard single-body configurations are inappropriate.

Air Valve Placement

In general, large orifice "air and vacuum" valves need to be located at pipeline summits. These locations are usually good collectors of air during pipeline operation. Small orifice "air release" valves should likewise be located at pipeline highpoints and at regular intervals, usually not exceeding 1,500 ft on long, flat slopes. Also, locate large orifice "air and vacuum" valves where negative pressures are possible during operations or draining, such as at the top of a steep downslope. Locating these valves near the operating hydraulic grade line (HGL) should be avoided. Fluctuations in pipeline pressure resulting in the HGL approaching the actual elevation of the valve can cause these valves to open and close rapidly and violently.

Air Valve Design

American Water Works Association's Manual M51 presents detailed guidelines for selecting, locating, and installing air valves in water and wastewater pipeline applications. Per AWWA Manual M51, the design of orifice size is affected by the pipe's internal pressure, the shape of the valve orifice, and the difficulty in determining the amount of air that may need to be vented. The manual provides a sizing method but emphasizes that the valve manufacturer should be consulted to verify the accuracy of assumptions made in the design process.

REFERENCES

American National Standards Institute/American Water Works Association (ANSI/AWWA). 201[illegible]. C900. Standard for Polyvinyl Chloride (PVC) Pressure Pipe and Fabricated Fittings, 4 In. Through 60 In. (100 mm Through 1,500 mm). Denver, Colo.: AWWA.

ANSI/AWWA. 2016. C909. Standard for Molecularly Oriented Polyvinyl Chloride (PVCO) Pressure Pipe, 4 In. (100 mm) and Larger. Denver, Colo.: AWWA.

ASTM International (ASTM). 2015. D1598. Standard Test Method for Time-to-Failure of Plastic Pipe Under Constant Internal Pressure. West Conshohocken, Pa.: ASTM International.

ASTM. 2014. D1599. Standard Test Method for Resistance to Short-Time Hydraulic Pressure of Plastic Pipe, Tubing, and Fittings. West Conshohocken, Pa.: ASTM International.

ASTM. 2015. D2241. Standard Specification for Poly (Vinyl Chloride) (PVC) Pressure Rated Pipe (SDR Series). West Conshohocken, Pa.: ASTM International.

ASTM. 2013. D2837. Standard Test Method for Obtaining Hydrostatic Design Basis for Thermoplastic Pipe Materials. West Conshohocken, Pa.: ASTM International.

ASTM. 2011. D3139. Standard Specification for Joints for Plastic Pressure Pipes Using Flexible Elastomeric Seals. West Conshohocken, Pa.: ASTM International.

Hucks, R.T. 1981. Changes in Strength of Pressurized PVC pipe with Time. *Journal AWWA.*

Jeffrey, J.D.; Moser, A.P.; and Folkman, S.L. 2004. *Long-Term Cyclic Testing of PVC Pipe.* Utah State University, College of Engineering, Logan, UT.

Killeen, N.D. and Schaul, J.S. 1964. Method of Determining Hydrostatic Design Stresses for PVC Pressure Pipe. *Interpace Technical Journal*, 1:1:17.

Malekpour, A.; Karney, B.; St. Aubin, R.; Martino, P.; and Gill, L. 2015. Exploring the Sensitivity of Fatigue Analysis with Regard to Design, Parameters in PVC Pipes Subjected to Cyclic Transient Pressures, 13th Computer Control for Water Industry Conference.

Moser, A.P. and Kellogg, K.G. 1994. Evaluation of Polyvinyl Chloride (PVC) Pipe Performance.

The Nature of Hydrostatic Stress Rupture Curves. 2005. PPI Technical Note PPI-TN 7. Plastics Pipe Institute, Irving, Tex.

Reinhart, F.W. 1973. Long-Term Hydrostatic Strengths of Thermoplastic Pipe. Proc. 4th American Gas Association Plastic Pipe Symposium, Arlington, Va.

Reinhart, F.W. 1961. Long-Term Working Stress of Thermoplastic Pipe. *SPE Jour.*, 17:8:75.

Streeter, V.L. and Wylie, E.V. 1967. *Hydraulic Transients*. New York: McGraw-Hill.

Uni-Bell Handbook of PVC Pipe—Design and Construction. 2012 (5th ed.). Dallas, Tex.: Uni-Bell PVC Pipe Association.

Watters, G.Z. 1971. The Behavior of PVC Pipe Under the Action of Water Hammer Pressure Waves. Utah State University, Utah Water Research Laboratory Report, PRWG-93.

Wilging, R.C. 1974. Stress Rupture Testing of PVC Pipe. *Modern Plastics*, 57:10:90.

This page intentionally blank.

Chapter 6

Receiving, Storage, and Handling

This chapter details procedures for receiving PVC and PVCO pipe, including inspection and unloading and recommended practices for subsequent storage and handling.

RECEIVING

When receiving PVC/PVCO pipe at the job site, the contractor or purchaser should follow the procedures outlined in this chapter.

Inspection

Each pipe shipment should be inventoried and inspected upon arrival. The carrier has the responsibility to deliver the shipment in good condition. The receiver has the responsibility to ensure that there has been no loss or damage. The records that accompany each shipment should provide a complete list of all items shipped.

The following procedures for acceptance of delivery are recommended:

- Examine the load. If the load is intact, inspection while unloading should be sufficient to ensure that the pipe has arrived in good condition.
- Some manufacturers cap the pipe ends to prevent contaminants from entering the pipe.
- If caps were installed at the factory, these should remain securely attached during storage to minimize contaminants entering the pipe.
- If the load has shifted, has broken packaging, or shows rough treatment, then each pipe length should be inspected for damage.

- If smaller-diameter pipe is "nested" in larger-diameter pipe, those pipes should be inspected carefully to ensure that no damage has resulted from shifting during shipment.
- Quantities of each item should be checked against shipping records (pipe, fittings, lubricant, etc.).
- Any damaged or missing items on the delivery receipt should be noted and recorded.
- Notify carrier immediately of any damage or loss and file a claim in accordance with the carrier's instructions.
- Damaged material(s) should be held and not discarded. It is the carrier's responsibility to provide information regarding the procedures that are to be followed.
- Replacements for shortages and damaged materials are normally not shipped unless requested. If replacement material is wanted, it needs to be reordered from the manufacturer, the distributor, or their authorized representative.
- Dark marks on the spigot ends of pipe are not a cause for concern. Each standard length of pipe is routinely hydrotested prior to shipping, and the gasket can leave dark streaks or marks on the spigot or bell end. The hydrotest is discussed in Chapter 2, Assurance Testing.

Unloading

The means by which PVC/PVCO pipe is unloaded in the field is the decision and responsibility of the receiver. Preferred unloading is in package units using mechanical equipment; however, the pipe can be unloaded individually by hand or mechanically.

When unloading package units, the following instructions should be followed.

- Remove restraints from the top unit loads. These may be either fabric or steel straps, ropes, or chains with padding for protection.
- If there are boards across the top and down the sides of the load that are not part of the pipe packaging, they should be removed.
- Use a forklift (or front-end loader equipped with forks) to remove the uppermost units from the truck one at a time. Remove units on the rear end of the truck bed first. Do not run the forks too far under units, as fork ends striking adjacent units may cause damage.
- If a forklift is not available, then a spreader bar may be used with fabric straps that are capable of handling the load. The straps should be placed at approximately the one-third points of the pipe or pallet and looped under the load. Cables also may be used if cushioned with rubber-hose sleeves or other material to prevent abrasion of the pipe.
- Follow supplier unloading instructions for fusible PVC. The lengths are longer, nominally 40 ft or more for this pipe, and may require more space or larger equipment. Bundles or pallets will tend to be heavier given the longer length. Ensure that properly sized equipment is used. These guidelines would also apply to any PVC/PVCO pipe having longer lengths.
- To unload lower units, repeat the described unloading process.

CAUTION:

- During removal and handling, be sure that the units do not strike anything.
- Severe impact could cause damage, particularly during cold weather.

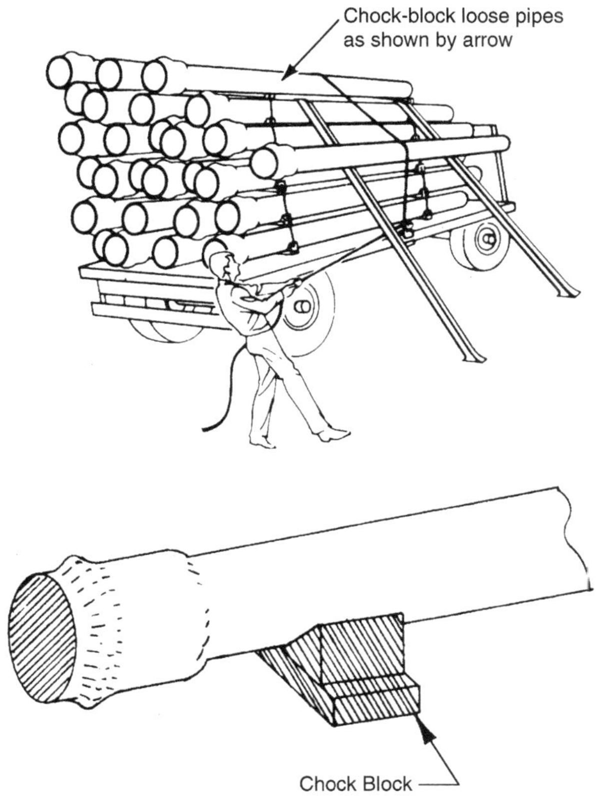

Figure 6-1 Chock block

- Do not handle units with individual chains or single cables, even if padded.
- Do not attach cables to unit frames or banding for lifting.
- Pipe package units should be stored and placed on level ground. Package units should not be stacked more than 8 ft high and should be protected by dunnage in the same way that they were protected while loaded on the truck.
- Follow local regulations.

If unloading equipment is not available, the pipe may be unloaded by removing individual pieces by hand. However, care should be taken to ensure that the pipe is not dropped or damaged (see Figure 6-1).

STORAGE

Damage to piping products can occur during storage. The following procedures and practices are recommended to prevent damage to PVC/PVCO pipe during storage.

- Regardless of the joining system (belled-end, fusible [plain-end], restrained-joint grooved end, etc.), the pipe should, when possible, be stored at the job site in the unit packages provided for shipment by the manufacturer. For pipes with belled ends (normally arranged alternately with pipe spigots), caution should be exercised to avoid compression, damage, or deformation of the bells.

- When unit packages of PVC/PVCO pipe are stacked, the weight of the upper units should not cause deformation to pipe in lower units. If pipe deformation is observed in the lower units, reduce the number of unit packages stacked above.
- PVC/PVCO pipe unit packages should be supported by racks or dunnage to prevent damage to the bottom during storage. Supports should be spaced to prevent pipe bending.
- Exposure to direct sunlight may result in frosting or chalking of the pipe. Chalking/frosting is an aesthetic rather than a performance issue. See the section in Chapter 1, "Environmental Effects, Weathering Resistance," for a discussion on chalking/frosting.
- When two or more years of exposure to direct sunlight is unavoidable or anticipated, PVC/PVCO pipe should be covered with an opaque material. Adequate air circulation above and around the pipe should be provided to prevent excessive heat accumulation (see "Environmental Effects, Thermal Effects" in Chapter 1).
- PVC/PVCO pipe should not be stored close to heat sources or hot objects, such as heaters, boilers, steam lines, or engine exhaust.
  - When unit packages of PVC/PVCO pipe are stacked, ensure that the height of the stack does not result in instability, which could cause stack collapse with resultant pipe damage or personal injury.
  - The interior and all sealing surfaces of pipe, fittings, and other appurtenances should be kept free of dirt and foreign matter.
  - All pipe and fittings gaskets should be protected from excessive exposure to heat, direct sunlight, oil, grease, and ozone.

REFERENCE

Uni-Bell PVC Pipe Association. *Uni-Bell Handbook of PVC Pipe—Design and Construction*. 2012 (5th ed.). Dallas, Tex.: Uni-Bell PVC Pipe Association.

Chapter 7

Installation in Trenches and Embankments

This chapter discusses recommendations for the installation of PVC and PVCO transmission and distribution water systems. These recommendations are subject to modifications for site-specific installation conditions and contractual obligations. Proper installation procedures should be based on a wide range of considerations, many of which are discussed in this chapter.

Installation is an important aspect of any pipe project. As discussed in Chapter 4, PVC and PVCO pressure pipe is flexible and engages lateral soil support to limit deflection. The soil support is dependent on the properties of the soil and quality of the installation.

Additional information on buried installation of PVC and PVCO pipe can be found in

- ASTM F1668, *Construction Procedures for Buried Plastic Pipe*,
- Uni-Bell UNI-PUB-9, *Installation Guide for PVC Pressure Pipe*, and
- AWWA C605, *Underground Installation of PVC Pressure Pipe and Fittings for Water*.

Additional information on soil compaction can be found in

- ASTM Manual 70, *Quality Control of Soil Compaction Using ASTM Standards* and
- *Pipeline Installation 2.0* (Howard, 2015)

ALIGNMENT AND GRADE

For most projects, the principal reference points for the specified pipeline alignment are provided by the engineer or owner. The principal reference points should be based on a permanent datum. The horizontal layout should be tied to a coordinate system in addition to property lines. Vertical control should be based on geodetic elevations. Traverses should be closed for horizontal control, and benchmarks should be checked for vertical control.

The transfer of the line and grade of the excavation work from control points established by the engineer is usually the responsibility of the contractor. This is a critical part of the project and must be checked thoroughly.

All pipe should be laid to and maintained at the established lines and grades. Fittings, valves, air-release valves, hydrants, etc., should be installed at the required locations with the valve and hydrant stems plumb.

Before work on the project begins, the entire site should be subjected to a thorough reconnaissance survey. The purpose of this survey is to identify potential problems during construction, which may include size and operating space requirements of equipment, intensity of traffic, and location of trees and overhead wires. All subsurface utilities, pipelines, and culverts need to be located during the planning stages to avoid construction delays due to location conflicts.

INSTALLATION IN TRENCHES

Most water mains are installed in a trench, which is a relatively narrow excavation dug in undisturbed soil. Some pipe may be installed in embankment conditions. The most common and effective way to install pipe in an embankment condition is to first build the embankment and then excavate the trench. Therefore, this chapter will focus on trench construction.

Terminology

The language used in the design, in the contract, and at the construction site should be clear and consistent. Figure 7-1 illustrates the common terminology and zones of a trench installation. The definitions of these terms are provided in Chapter 4 and are illustrated again here for convenience.

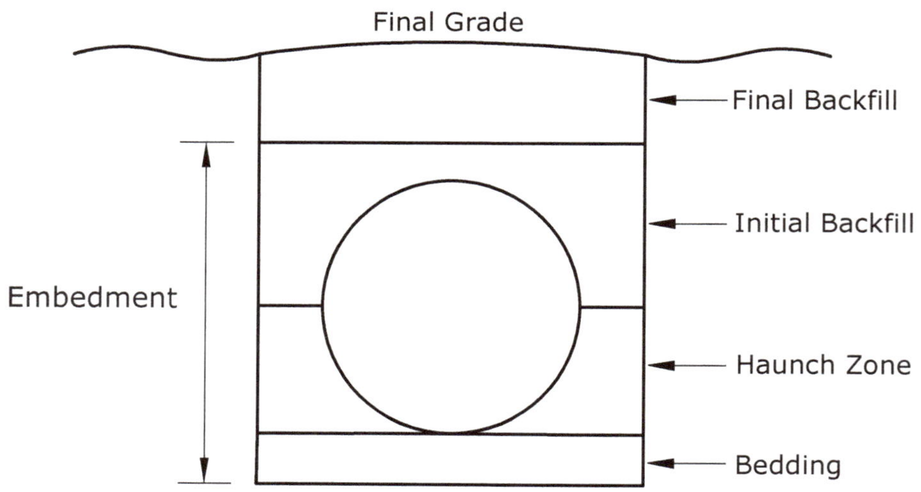

Figure 7-1 Trench construction and terminology

Trench Width

The trench width at the ground surface may vary depending on depth, nature of the in situ soils, and proximity of adjacent structures. If compaction of pipe embedment is necessary, the minimum clearance from the trench wall to any portion of the pipe, bell, or appurtenance should be at least 9 in. (230 mm). In poor soils, the width may need to be increased to properly support the pipe; see Chapter 4 for required trench width using composite *E′* based on the engineered installation design. The contractor may elect to use various methods to excavate and to support the trench in accordance with regulatory requirements. Regardless of the method, if compaction is required, the trench width must always provide sufficient room between the trench wall and the pipe for proper compaction.

When compacted embedment is required and the trench is excavated in a recently constructed embankment, the stiffness of the embankment soil (within two or more pipe diameters on each side of the pipe) should equal, or be greater than, that required for the embedment unless a composite stiffness is used in the design (see Chapter 4).

The trench backfill or embankment must be properly compacted if settlement at the surface needs to be prevented.

A trench protection system may be required by law in locations where poor soils are present during construction, in congested areas that have many utilities that require support during construction, or in confined working areas.

The contractor should be responsible for designing, installing, and removing any trench support systems. The trench support system should be designed by a professional engineer. The geotechnical report may provide parameters that can be used for the design of the trench support systems. National, state, and local safety requirements must be followed, and the people involved in the construction supervision and administration need to be familiar with these requirements.

AWWA C605 Trench Types

Figure 7-2 shows the trench types as illustrated in ANSI/AWWA C605.

BASIC AND ENGINEERED INSTALLATIONS

Alternatively, *basic* or *engineered* installations may be used. The requirements for basic and engineered installations are outlined in Chapter 4. Installation considerations vary considerably for the two types.

Basic Installation

Flexible pipe with sufficient stiffness may be installed with minimal soil support. Pipe can be laid directly on the trench bottom. The soil excavated from the trench can often be used for the pipe embedment.

A basic installation (as described in Chapter 4 and repeated here) can be used for the following conditions:

- Nominal pipe size of 24 in. or less.
- DRs equal to or less than 25.
- Depths of cover 15 ft or less.
- Natural groundwater below pipe.
- For depths of cover with less than 6 ft, no live load or surcharge load.
- Final backfill compaction is not required.

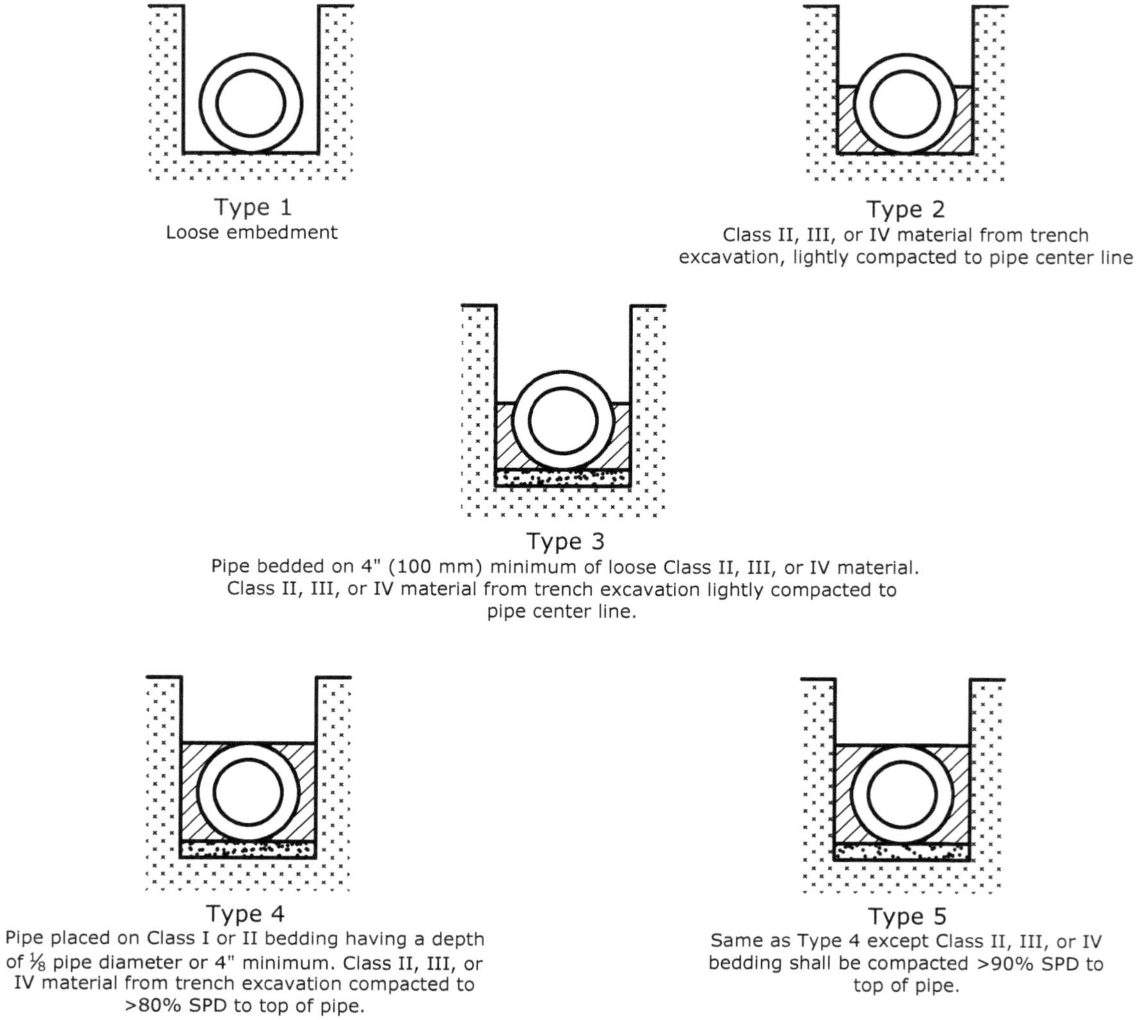

Notes:

1. See Table 4-1 for descriptions of Class I, II, III, IV, and V soils.
2. Do not use Class V soils for bedding or embedment. Class V soils may be used for backfill, where appropriate.
3. SPD is Sandard Proctor Density, AASHTO T-99 or ASTM D698.

Figure 7-2 Standard trench types for PVC and PVCO pipe

- Embedment soil *E′* will be at least 200 psi (as discussed in Chapter 4 and later in this chapter).
- Trench foundation and trench walls are stable and have a minimum unconfined compressive strength of 5 psi, an N value of at least 5 from the standard penetration test, or a *E′* of at least 400 psi (as discussed later).
- The foundation does not consist of expansive clays, collapsing soils, or landfill.
- The soils in the foundation and used for the embedment do not contain rock particles larger than the allowable maximum particle size.

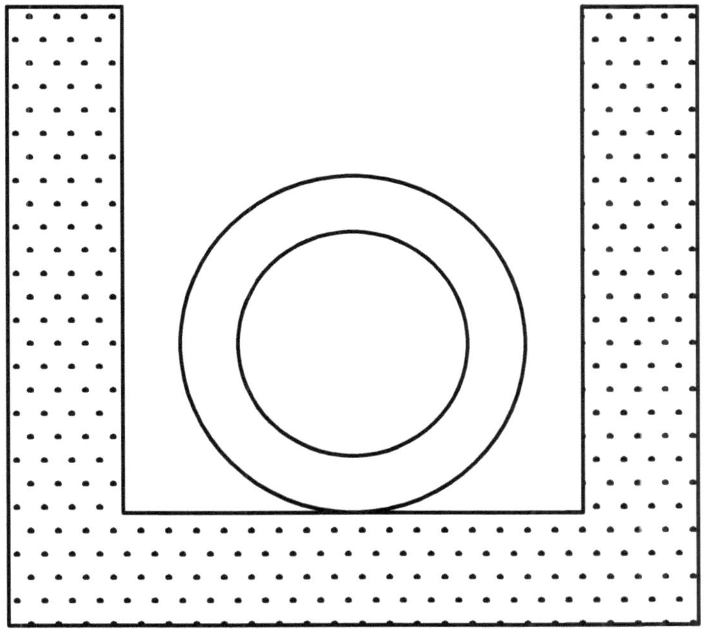

Source: Adapted from US Bureau of Reclamation

Figure 7-3 Basic installation for PVC and PVCO pipe

Figure 7-3 represents a basic installation.

Where crossings occur along a basic installation system, an engineered installation should be used for the crossing portion of the pipeline. Crossings include roadways, waterways, pipelines, and other buried utilities.

Engineered Installation

When the pipe stiffness is not sufficient to carry the anticipated loads, basic installation is not appropriate. Also, where the pipe crosses under a roadway, waterway, or another pipeline, an engineered installation may be required for those portions of pipe.

An engineered installation design considers the trench wall support, the effects of groundwater, selection of embedment material, soil compaction, time before pipeline is pressurized, live load, and surcharge load. Chapter 4 describes the conditions for these concerns. Installation may require imported embedment material, a bedding for the pipe, soil testing, and more stringent inspection.

Figure 7-4 represents an engineered installation.

Trench types 2, 3, 4, and 5, as shown in Figure 7-2, are standard engineered embedments for PVC and PVCO pressure pipe.

These engineered installation types may be modified to satisfy the specific requirements covered in Chapter 4. The design objectives depend on proper installation. During construction, any field deviations from the specified installation must be approved by the design engineer.

SOIL CLASSES AND TREATMENT

The soil classes, as defined by the Unified Soil Classification System (USCS), Class I to Class V, in Table 7-1 are in descending order of stiffness when the soil is compacted. Class I and Class II soils are usually considered cohesionless and are best compacted using vibration. Class III and Class IV are usually considered cohesive and are best compacted

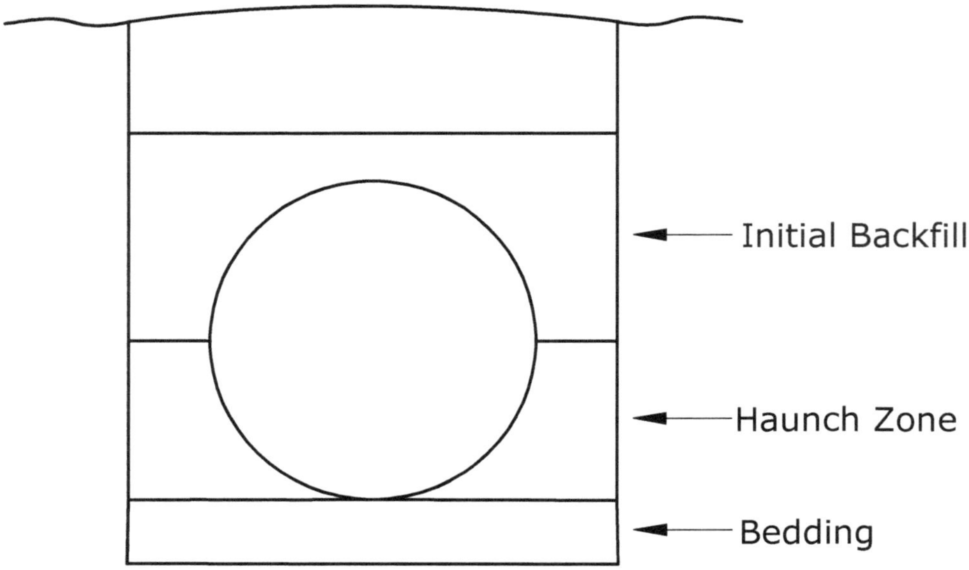

Source: Adapted from US Bureau of Reclamation

Figure 7-4 Engineered installation for PVC and PVCO pipe

Table 7-1 Soil classes for pipe installation

Soil Class	Description	USCS Symbol
Class I	Crushed rock	
Class II	Clean, coarse grained soils	GW GP SW SP
Class III	Coarse grained soils with fines	GM GC SM SC
	Sandy or gravelly fine grained soils	ML CL
Class IV	Fine-grained soils	ML CL
Class V	Fine-grained soils, organic soils	MH CH OL OH Pt

Source: Howard, A. 2015

with pressure, impact, or kneading. Class V soils are considered cohesive but are not recommended for use in pipe installation. A more complete description of the soil classes is included in Chapter 4, Table 4-1.

Excavating

The trench must be excavated to the approximate grade of the pipe with a width as shown in the construction drawings. The trench depth, native soil materials, presence of rock, water table, location (rural or urban), other utilities, and traffic requirements must all be considered before installation and managed during installation.

The depth of the excavation required to install the pipe can be determined from the design drawings. Some additional excavation may be required under the pipe because of unsuitable foundation material (see later discussion).

The native soil materials and the presence of rock at the site can be anticipated from the geotechnical report prepared for the project. This report should be kept onsite at all times for reference. The actual soils encountered should be regularly recorded in the site diary to document the project and any changed conditions.

Dewatering

Dewatering or pre-drainage of the trench site is usually carried out by vacuum well-points, deep wells, or eductor wells. The function of dewatering is to efficiently remove only enough water from an aquifer to facilitate underground construction. The dewatering system should lower the water level in the trench during installation of the pipe to a level below the bottom of the pipe. The choice of dewatering method depends on aquifer characteristics, depth of required drawdown, site geometry, and cost.

Vacuum well-point system. A vacuum well-point system relies on atmospheric pressure to remove the groundwater upward to the ground surface where it can be collected or discharged. Well-point systems are most suitable in shallow aquifers where the water level needs to be lowered no more than 15 or 20 ft. As the distance increases from the well-point, the amount of drawdown will depend on the aquifer characteristics. Well-point spacing is a function of the depth of required drawdown, soil permeability, and the presence of impermeable layers.

The well-point system includes a vacuum unit, which is used to create low air pressure in the header well-point systems. As a result of the low pressure at the well-point, atmospheric pressure forces water up to the surface. The groundwater is then collected at the surface in a separation tank and discharged to an appropriate location. Well-point systems have decreasing capability at higher elevations.

Deep-well system. Deep wells are used where there is a large wetted volume below the excavation level and pervious soils exist, resulting in high well yields. Deep wells are the simplest and oldest method of removing water from the ground. In a deep-well system, the groundwater flows by gravity to a well, where it is removed by pumping. The pumping unit is usually a submersible turbine pump. If a large amount of water is to be pumped, vertical line shaft pumps with the motors at ground surface are used.

Eductor system. The eductor system is usually used for dewatering fine-grained soil where the depth of dewatering is greater than 20 ft (6.6 m).

An eductor system can employ either a single or double pipe design. Single pipe systems consist of a large-diameter pipe that forms the well casing with a small inner pipe that forms the return line. Water is pumped under high pressure through the annular space between the two pipes and is forced through a nozzle and venturi. This causes low pressure to develop around the nozzle. The low pressure will draw water to the well and up the return line. Typical operating pressures are 7–10 atmospheres on the pressure side.

Eductor systems typically have a limited quantity of water that can be pumped, usually ranging from 3 to 11 gpm (0.19 to 0.69 L/s). An advantage of the eductor system is that the vacuum is developed directly at the well screen; with a properly installed seal above the sand pack around the well-point, a considerable vacuum can be transmitted to the soil.

Dewatering systems—in general. Water in a trench can also be controlled by using a drainage rock underneath the pipe, sometimes with drain pipes, to carry the water to sumps where the water can be pumped out.

The general contractor and his dewatering subcontractor are usually assigned the responsibility of deciding upon the need, design, operation, maintenance, and removal of the dewatering systems. The owner is usually responsible for providing adequate information on the ground conditions. The information should include detailed bore-hole logs, where groundwater was encountered during drilling, water level in piezometers sealed into specific water-bearing stratum, and size distribution curves of the water-bearing soils.

Stockpiling Excavated Materials

Excavated material should be stockpiled in a manner that will not endanger the work activities or obstruct sidewalks, driveways, or public safety devices. The stockpiled material must not damage trees and lawns.

Stored material should be kept back from the wall of the trench to prevent loading, which could lead to a cave-in.

Fire hydrants under pressure, valve pit covers, valve boxes, curb stop boxes, and fire and police call boxes and other utility controls should remain unobstructed and accessible. Gutters and ditches shall remain clear unless other arrangements have been made for drainage. Natural watercourses should not be obstructed, and an attempt shall be made to try to anticipate what might happen during a rainstorm or extended exposure to the elements.

Trench Foundation

The foundation, as shown in Figure 7-1, is the native material (or compacted embankment) in the bottom of the trench. The trench foundation should provide a firm, stable, and uniform support for the full length of the pipe or pipe bedding.

Where a soft trench foundation is encountered, over excavation and backfilling with select material may be necessary for stabilization. The depth of the stabilization should be determined by tests and observations, and a geotechnical consultant may be required.

In wet conditions, Class I or II material used for stabilization should be wrapped in a geotextile to prevent migration of soil particles from the foundation material. In cases where a trench foundation cannot be stabilized and where intermittent areas of unequal settlement are anticipated, special foundations for the pipe may be necessary.

The trench foundation should be cleaned of all loose or projecting rocks before placement of the bedding material as they may cause point loading on the pipe that could lead to pipe failure.

If blasting is required to remove rock, such work must be performed by blasting specialists.

Before blasting is carried out on any site, a careful examination, preferably including photographs of all buildings and structures liable to be within the range of perceptible vibration from the source, should be carried out. If this pre-blast survey is carefully made, all existing cracks and defective conditions can be recorded, and their number and locations can be brought to the attention of the owner.

PIPE BEDDING

Bedding Type and Material

Bedding (Figure 7-1) for the pipe is not required for the basic installation (Figure 7-2, Type I, and Figure 7-3) provided the foundation is not rock and the foundation soil does not contain rock particles larger than the allowable maximum particle size specified by the design engineer.

For engineered installations, bedding provides uniform support under the pipe. The bedding thickness should be at least the pipe diameter divided by 12 with a minimum of 4 in. The bedding soil should be evenly spread across the trench bottom to the correct grade and need not be compacted. The bedding, as well as the rest of the embedment material, should not contain any rocks larger than the allowable maximum particle size specified by the design engineer.

If the foundation is rock, the bedding thickness should be increased 2 in. with a minimum of 6 in. to minimize point loading. The thickness should be measured from the peaks or raised portions of the rock foundation surface.

Trimming of the excavated trench bottom to fit the pipe barrel is not essential and is very difficult to accomplish with any accuracy. Pipe can be placed directly on the bedding.

In wet conditions, embedment soils that are Class I or II may need to be wrapped in a geotextile to prevent migration of the particles into the bedding material from the foundation, trench walls, or embedment.

Laying of Pipe

The full length of the pipe barrel should be in contact with the foundation (basic installation) or the bedding (engineered foundation) in order to provide uniform support. Bell holes should be dug to prevent point bearing on the bottom of the bell. There should be clearance of an inch or more underneath the pipe bells.

Pipe and accessories should be inspected for defects, damage, and cleanliness before they are lowered into the trench. Any defective or damaged material should be repaired or replaced. All foreign matter or dirt should be removed from the interior of the pipe and accessories before they are lowered into the trench.

All pipe, fittings, valves, hydrants, and accessories should be carefully lowered to prevent damage. Pipe and accessories should never be dropped or dumped into the trench.

When pipe laying is not in progress, open ends of installed pipe should be capped or otherwise closed to prevent entrance of trench water, dirt, foreign matter, or small animals into the line. Whenever water is prevented from entering the pipe, enough backfill should be placed on the pipe to prevent floating.

PIPE JOINTS

Gasket Joint Assembly

The assembly of PVC and PVCO pipe requires the careful adherence to the proper joint assembly procedures outlined in the contract documents. The following is a suggested procedure.

1. Confirm that the bell and gasket are free from any foreign material that could interfere with the proper assembly of the pipe joint. Some gaskets are restrained in the bell and should not be removed. Contact the manufacturer for specific information.
2. Confirm that the pipe spigot end is clean. Wipe with a clean, dry cloth around the entire circumference from the pipe end to about 1 in. (25 mm) beyond the reference mark.
3. Lubricate the spigot end of the pipe using a lubricant and method of application recommended by the pipe manufacturer. The entire circumference should be lubricated, especially the beveled end of the spigot. After the spigot end has been lubricated, it must be kept clean and free of dirt and sand. If dirt or sand adhere to the lubricated end, the spigot must be wiped clean and re-lubricated.
4. Lubricate the sealing surface of the gasket in the bell. Do not lubricate the underside of the gasket or the gasket groove in the bell because the lubrication could cause gasket displacement.

5. If the system is for potable water, the lubricant must be approved for potable water service. Do not use an unapproved lubricant that may harbor bacteria or damage the gaskets or pipe.
6. The bell or coupling shall be in straight alignment with the pipe spigot end before and during joint assembly. Push the lubricated spigot end into the bell or coupling beyond the gasket. The bar and block method of assembly is recommended as the worker can feel the amount of force being used and whether the joint slides together smoothly. Larger pipe may require mechanical assistance to apply sufficient force to assemble the joint. However, it is critical that the spigot is not installed beyond the insertion line, which may lead to additional or irreparable damage due to mechanical assistance. Non-gouging lifting straps are only permitted to install and support pipe.

   Pipe spigot ends are pre-marked with one or two circumferential insertion lines. These lines reference how far the spigot shall be inserted into the adjoining PVC pipe bell. Field-cut pipe spigot ends shall be marked to match the manufacturer's insertion line(s) and beveled to match factory-supplied bevel.

   In the case of single-insertion line pipe, spigots should be inserted into the adjoining bell or fitting until the edge of the bell or fitting is flush with the insertion line. After assembly, the insertion line shall remain visible and be nearly flush with the lip of the adjoining pipe bell or fitting. Joints assembled beyond the insertion line shall be considered over-assembled and may result in damaging stresses or leakage. Joints not assembled up to the single insertion line may not seal properly.

   In the case of dual-insertion line pipe, spigots should be inserted into the adjoining bell or fitting until the edge of the bell or fitting is between the two insertion lines. After assembly, one insertion line shall be hidden inside the joining while the other shall remain visible. Joints assembled beyond the second insertion line shall be considered over-assembled and may result in damaging stresses or leakage. Joints not assembled beyond the first insertion line may not seal properly.
7. If undue resistance to insertion of the spigot end is encountered or the reference mark does not reach the flush position, disassemble the joint and check the position of the gasket. If it is twisted or pushed out of its seat, replace the pipe, or the gasket, in accordance with the pipe manufacturer's recommendations. If the gasket is not out of position, measure the distance between the reference mark and the spigot end and check it against the correct values provided by the pipe manufacturer.

Pipe Joint Offset

The gasketed joints of PVC and PVCO pipe are designed to accommodate minor changes in grade and alignment. However, specified changes in alignment and grade should be accomplished with deflection couplers or fittings.

As a practical matter, PVC joints may be offset slightly until the point where the outside diameter of the spigot contacts the inside diameter of the bell. The maximum allowable axial joint deflection in degrees (for example, joint $D \leq 1°$) may be found on the pipe marking. Each pipe joint shall be checked during installation to verify the axial joint deflection or offset is not exceeded prior to backfill. Otherwise, the pipe manufacturer should be contacted for the amount of offset allowed. (See Joint Offset in Chapter 12.)

PIPE CUTTING AND BENDING

Pipe Cutting

To set a fitting, valve, or other appurtenance in a specific location, it may be necessary to cut the pipe. A square cut is essential to ensure proper assembly. Both portions of pipe on either side of the cut line should be supported from below such that neither portion of pipe pulls at the other while it is being cut. AWWA C900 and C909 pipe can be easily cut with a fine-tooth hacksaw, handsaw, or a portable power saw with a steel blade or abrasive discs. Chainsaws shall not be used to cut PVC or PVCO pipe. Before the pipe is cut, it should be marked around its entire circumference to assist in making a square cut.

Use a factory-finished pipe end as a guide to determine the angle and length of bevel. After cutting and beveling, at least one reference mark must be made on the spigot end if it will be inserted into a PVC or PVCO fitting or pipe bell. The reference mark(s) can be properly located by either measuring the depth of the PVC or PVCO fitting bell and subtracting an inch or, if it will be joined to another PVC or PVCO pipe, using a factory-marked end of that PVC or PVCO pipe as a guide. The reference point location(s) may also be provided by the pipe or fitting manufacturer.

Installers should be aware that when assembling two different brands of PVC or PVCO pipe, the reference line position may differ. In this case, the correct reference line for the spigot of Brand "X" is obtained by measuring the insertion depth of an uncut length of Brand "Y."

For iron fittings (push-on or mechanical joints), remove all but ¼ in. (6 mm) of the factory-made bevel from the spigot end to ensure full contact of the spigot outside diameter with the gasket. The PVC or PVCO pipe spigots should be fully inserted (bottomed-out or "homed") in the iron fittings.

Pipe Bending

PVC and PVCO pipe does offer the advantage of longitudinal bending to accomplish changes of direction. If the pipe is to be bent, precautions must be taken to ensure straight alignment in the joint. Consequently, longitudinal bending with gasket joints is limited to the smaller pipe diameters as listed in Chapter 12. Also, the pipe must meet minimum bending radius requirements at every point (see Performance Limits in Longitudinal Bending in Chapter 12).

HAUNCH ZONE AND INITIAL BACKFILL

For the basic installation, the haunch zone and initial backfill is typically the material excavated from the trench and dumped back in around the pipe. However, care must be taken to not shove the pipe out of alignment, to prevent severe impact on the top of the pipe, and to ascertain that there are no rock particles that are larger than the allowable maximum particle size specified by the design engineer.

For the engineered installation, pipe deflection is resisted, in part, by the stiffness of the haunch zone and initial backfill soil. Therefore, this portion of the embedment must be properly constructed. Close inspection and testing is necessary to ensure that the installation will reflect the conditions assumed in the design of the pipe. The embedment soil type and the compaction have a direct bearing on the amount of support it will provide to the pipe. Density tests should be conducted to measure the percent compaction of the haunch zone and initial backfill.

The maximum embedment particle size should not exceed ¾ in. (20 mm) for angular rock or 1½ in. (40 mm) for rounded rock, and should be no larger than 10 percent of the

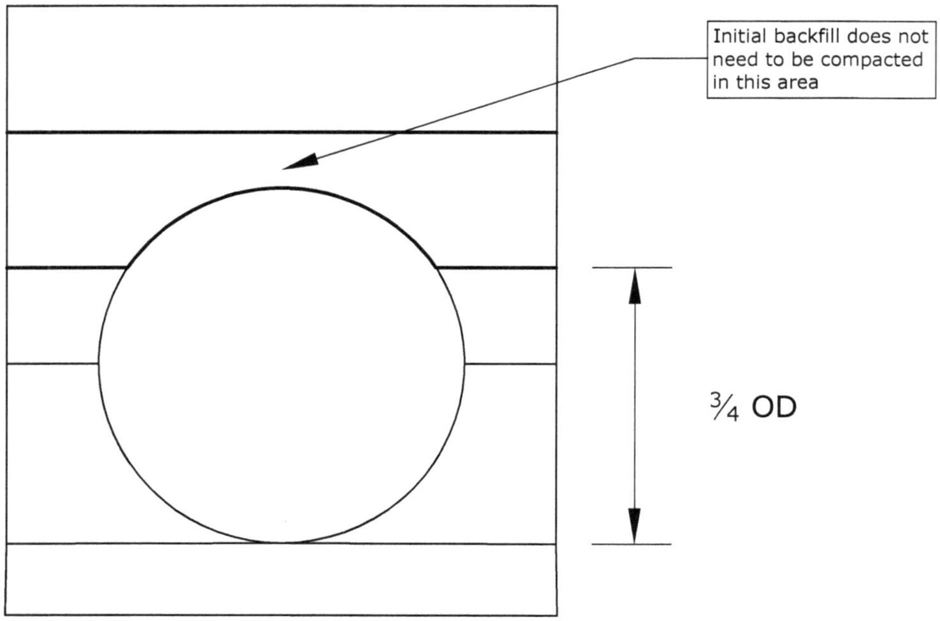

Source: Adapted from US Bureau of Reclamation

Figure 7-5 Diagram showing lower boundary of initial backfill

pipe diameter for smaller pipes. The embedment soil should not contain any organic matter, debris, or frozen lumps.

For the engineered installation, the pipe haunch material must be compacted to the specified level to provide support of the pipe. Compaction of the initial backfill should be limited to a level no higher than three-quarters of the pipe diameter above the pipe invert, as shown in Figure 7-5. The initial backfill directly over the pipe (at least 12 in. above pipe) need not be compacted and is best left uncompacted.

SOIL COMPACTION

The soil support for the pipe is dependent on the degree of compaction, referred to as *percent compaction*. Percent compaction is defined by ASTM D653 as the ratio of the field compaction to the laboratory maximum density, expressed as a percentage. The field compaction can be measured by in-place density tests such as sand cone or nuclear gauge. For soil groups III, IV, or V, the laboratory maximum density is determined using the standard Proctor compaction test ASTM D698. For soil groups I or II, the laboratory maximum density is determined using a vibratory compaction test ASTM D7382 or D4253. ASTM D7382 is a newer procedure using a vibratory hammer to obtain a maximum density and is considered more reliable than D4253.

References to the percent compaction of soils in this manual will be as recommended in ASTM D653. The percent of the maximum density of the soil is followed by the ASTM test procedure used to determine the maximum density. For example, 95 percent (D698) means that the in-place density should be equal to or higher than 95 percent of the maximum density obtained using D698. The references in this manual are based on standard Proctor density, which is ASTM D698. Therefore, plans and specifications should reference D698. Testing and inspection during construction is important to ensure that the installation meets the design criteria.

Final Backfill

Final backfill, as shown in Figure 7-1, can be pushed into place and may or may not require compaction. The degree of compaction usually depends on the location of pipe relative to roadways, railways, or other structures, and/or if settlement at the surface needs to be prevented (e.g., cultivated lands). Compacted backfill can be Class I, II, III, or IV material. Class V soil is not recommended for compacted backfill.

Within the final backfill area 1–3 ft from the pipe, the material should not contain any rocks larger than 6 in. The backfill should be free of trash, organic matter, and debris.

When compacting material over the pipe, place at least 12 in. (30 cm) of loose material over the pipe before compacting directly over the pipe with handheld or walk-behind compaction equipment. Do not use ride-on compaction equipment, until at least 3 ft (1 m) of material is placed over the pipe.

The entire trench must be properly backfilled and compacted if settlement at the surface needs to be prevented.

Flowable Fill

Flowable fill is a fluid mixture of cement, soil, and water that hardens into a material stronger than traditional compacted soil. Portland cement or Class C fly ash can be used as the cementitious material. Clean concrete sand or earth excavated from the trench can be used as the soil. Recycled concrete, recycled glass, foundry sand, and other waste products have been successfully used as the aggregate. Mixing equipment can be transit mix trucks, volumetric concrete mixers, or several types of trench side mixers. The flowable fill can be designed and installed so the pipe can be backfilled or paved over the next day. The pipe is supported toward each end on soil pads or sand bags so the mixture will flow readily under the pipe. The pipe typically has to be restrained in order to prevent floating. One advantage is that only the soil pads have to be adjusted for grade, not the entire bedding surface.

If soil or sand bags are not used as support, then the material supporting the pipe must be a weaker material than the flowable fill in order to prevent future point loads on the pipe.

Flowable fill can be used in place of any of the embedment soils shown in Figure 7-1. The 28-day compressive strength of the flowable fill should be between 50 and 100 psi, as determined by ASTM D4832 and have a spread of 8–12 in. as determined by ASTM D6103. The low strength provides a material that can be readily excavated in the future.

In some cases, particularly under streets, flowable fill can also be used as the final backfill over the pipe. This will replace compacting the backfill and minimize settlement.

PVC Fittings Installation

Following are some guidelines for proper installation of PVC fittings.

1. If rocky conditions exist at the trench bottom, a minimum layer of 4 in. (100 mm) of bedding should be placed below the fitting. Bell holes should be dug to ensure uniform loading on the fitting.
2. Class I, II, III, or IV backfill material should then be placed and may be compacted around the sides of the fitting. Compaction should be achieved using only tamping bars when within 6 in. (150 mm) of the fitting. Mechanical compactors may be used elsewhere.
3. Backfill material for the bedding, haunch, and initial backfill zones around fittings should not contain rock particles larger than 1½ in. (38 mm) in size.

4. Final backfill should be free of sharp objects, sticks, or boulders. Ride-on compaction equipment should never be used within 3 ft (1 m) of the PVC fitting.
5. Compaction is especially important on both sides of a branch of a PVC tee or wye to minimize lateral bending stresses. Thrust restraint design is covered in Chapter 11.
6. Design and installation of non-PVC appurtenances within PVC and PVCO piping systems should take into account the mass of the appurtenance. Additional foundation material and/or bedding compaction may be required to minimize differential settling, bending moments, point loads, etc., on adjacent pipe.

REFERENCES

American National Standards Institute/American Water Works Association (ANSI/AWWA). 2013. C605. Underground Installation of Polyvinyl Chloride (PVC) and Molecularly Oriented Polyvinyl Chloride (PVCO) Pressure Pipe. Denver, Colo.: AWWA.

ANSI/AWWA. 2016. C900. Polyvinyl Chloride (PVC) Pressure Pipe and Fabricated Fittings, 4 In. Through 60 In. (100 mm Through 1500 mm). Denver, Colo.: AWWA.

ANSI/AWWA 2016. C909. Molecularly Oriented Polyvinyl Chloride (PVCO) Pressure Pipe, 4 In. (100 mm) and Larger. Denver, Colo.: AWWA.

ASTM International. 2014. D653. Standard Terminology Relating to Soil, Rock, and Contained Fluids. West Conshohocken, PA.: ASTM International.

ASTM International. 2012. D698. Standard Test Methods for Laboratory Compaction Characteristics of Soil Using Standard Effort (12,400 ft-lbf/ ft^3 (600 kN-m/m^3)). West Conshohocken, Pa.: ASTM International.

ASTM International. 2012. D2774. Standard Practice for Underground Installation of Thermoplastic Pressure Piping. West Conshohocken, Pa.: ASTM International.

ASTM International. 2016. D4253. Standard Test Methods for Maximum Index Density and Unit Weight of Soils Using a Vibratory Table. West Conshohocken, Pa.: ASTM International.

ASTM International. 2016. D4832. Standard Test Methods for Preparation and Testing of Controlled Low Strength Material (CLSM) Test Cylinders. West Conshohocken, Pa.: ASTM International.

ASTM International. 2017. D6103. Standard Test Methods for Flow Consistency for Controlled Low Strength Material (CLSM). West Conshohocken, Pa.: ASTM International.

ASTM International. 2008. D7382. Standard Test Methods for Determination of Maximum Dry Unit Weight and Water Content Range for Effective Compaction of Granular Soils Using a Vibrating Hammer (Withdrawn 2017). West Conshohocken, Pa.: ASTM International.

ASTM International. 2016. F1668. Standard Guide for Construction Procedures for Buried Plastic Pipe. West Conshohocken, Pa.: ASTM International.

ASTM International. Manual 70. *Quality Control of Soil Compaction Using ASTM Standards*. West Conshohocken, Pa.: ASTM International.

Howard, A. 2015. *Pipeline Installation 2.0*. Harpers Ferry, W.Va.: Relativity Publishing.

Installation Guide for PVC Pressure Pipe. Dallas, Tex. Uni-Bell PVC Pipe Association.

Chapter 8

Testing and Maintenance

This chapter contains procedures for hydrostatic testing and for disinfection before PVC or PVCO potable water piping is placed in service. Procedures are also presented for thawing and locating PVC and PVCO piping. Installers should refer to the contract documents for the required testing procedures for new construction.

The use of water as the test medium is specifically recommended for water mains. While air testing at low pressures (up to 5 psi) is an accepted practice for gravity sewer installations, it is not an accepted or recommended method for testing for a pressurized water main. As with all pipe materials, the removal of air from the test section is a critical step in testing the line safely and eliminating pressure drop during testing caused by compressed air.

INITIAL TESTING

During the initial stages of the construction project, a short length of installed pipeline should be tested. This will allow the installer to verify that proper installation and joint assembly techniques are being employed.

The length of section to be tested will depend on the specific project; however, the length should be long enough to ensure a representative length is used with at least 20 pipe joints. To avoid additional costs caused by restrained joints and temporary blocking, it may be possible to adjust the initial test length to suit facilities already required for final installation.

The requirement for initial testing should be clearly stated in the contract documents to avoid a claim for delay of work. Some contractors may fail to see that the requirement for initial testing is beneficial to all involved in the project. The initial test pressure should be to the same hydraulic grade line as the final testing. The testing procedure is presented later in this chapter.

TIMING OF THE TESTING

To prevent floating or movement, the pipeline should be kept in a dewatered trench condition until the pipeline is filled with water. To prevent movement of the pipe resulting from the pressure test, sufficient backfill should be placed before filling the pipe with water and subsequent field testing. Where local conditions require that the trenches be backfilled immediately after the pipe has been laid, the testing may be completed after backfilling but before placement of the permanent surface.

At least seven days should elapse after the last concrete thrust or reaction blocking has been cast with normal (Type 1) Portland cement. The elapsed time may be reduced to three days with the use of a high-early-strength (Type III) Portland cement. Use of Type III cement is cautioned when PVC fittings are used due to heat of hydration.

Restrained joints must be assessed if testing is to be done immediately after trench backfilling. At this stage of installation, the full weight of the trench backfill may not have come to bear on the pipe. If this is the case, the frictional resistance at the time of the testing may be less than that anticipated by the designer. The designer of the water main should be consulted before the testing.

INITIAL CLEANING OF THE PIPELINE

The pipe must be cleaned before disinfecting the water main. Mud, sand, dirty water, or a variety of debris left in the water main can provide a growing medium for bacteria and will also serve to shield bacteria from contact with the chlorine solution. This can inhibit disinfection and possible delivery of contaminated water to consumers. While it is often assumed that water mains can be successfully cleaned by flushing or swabbing, some forms of contamination can prove resistant to those methods and require significant effort and resources to pass bacteriologic tests. To avoid difficulties with disinfection of the main, effort should be made to keep the interior of the pipe clean at all times.

Water mains should be filled slowly with potable water at a rate that allows air to leave the line at the same rate as the water entering the line. (Example: 2-in. air valve, $Q = 2.18$ ft^3/s.) All air should be vented from the pipe and fittings to prevent entrapment of air in the main. For flushing of the pipeline, it is suggested that sufficient flow in the system be created to cause the pipeline flow velocity to be equal to or greater than 3.0 ft/s. The duration of the initial flushing procedure should be continued until the discharge appears clean; however, the minimum duration should be based on a minimum of three changes of pipeline volume. The discharge of flushing water must be carefully managed to prevent damage to property or interruption to traffic.

Foam swabbing may be used in lieu of flushing. This procedure should be carefully assessed before undertaking the work.

TEST PREPARATION

Proper preparation for the hydrostatic test will help ensure that the test is conducted in a safe and efficient manner. It is critical that all facilities required to handle thrust restraints are properly installed. Thrust facilities must also include the trench backfill if restrained joints are used. Concrete thrust blocks must be allowed to develop the required strength.

To minimize the amount of makeup water needed and make the test more meaningful, the following procedures should be performed.

1. Close all outlets.
2. Properly tighten all bolts on water main appurtenances and flanges. All water main appurtenances, flanges, test hardware, and restraints shall be installed in accordance with the manufacturer's instructions before testing.

3. Release air from all high points.
4. Fill the main with test water from a low point.
5. Fill the line at a slow rate (maximum velocity of 1 ft/s [0.3 m/s]) to minimize air entrapment and potential surge pressures.
6. After filling the line, flush the line for a sufficient period to help move any remaining air to the release valves or other outlets. It may be necessary to repeat this step after allowing the line to sit to allow entrapped air to accumulate after flushing.
7. Check the operation of all pumping equipment and calibrate or certify gauges and meters before conducting test.

HYDROSTATIC TESTING

Hydrostatic testing is recommended to establish the integrity of the pipeline and to determine if the bell and spigot are joined properly. The test is usually conducted by the contractor and witnessed by the owner or the owner's representative.

The "test allowance" is the quantity of makeup water that must be supplied to the test section to maintain a given pressure. If the quantity of makeup water added is less than the predetermined test allowance, the pipeline is considered acceptable.

In addition to leakage, there are other factors that can increase the amount of makeup water needed to maintain the given test pressure. These factors include compressing trapped air, take-up of restraints, and temperature variations during testing. All visible leaks observed during the hydrostatic test should be repaired. The contract documents should include a test allowance; however, some authorities state that no makeup water should be permitted. Recommended test allowances for PVC and PVCO pipe installations are provided in this chapter.

TEST PRESSURE

To test the system, the hydrostatic pressure should be increased to at least the maximum operating pressure of the line. Many jurisdictions require that the system be subjected to pressures in the range of 120–150 percent of the maximum working pressure. The allowance over the maximum working pressure should be carefully considered. A high allowance can increase the cost of the system as a result of the size of thrust blocks, number of restrained joints, and the increased pressure rating of appurtenances such as flanges, valves, etc.

The required hydrostatic test pressure should be presented as a hydraulic grade line rather than a gauge pressure. This is particularly important in hilly terrain. A vertical rise of 100 ft (33 m) translates into approximately 43 psi (295 kPa). Thus, if the hydrostatic test pressure is identified as 150 psi (1,035 kPa), then to get a reading of 150 psi (1,035 kPa) at the top of a 100-ft-high (33-m) hill, the pressure would be 193 psi (1,330 kPa) at the bottom of the hill. Stating the test pressure in terms of hydraulic grade line makes it easier to understand the effects of the test pressure on the pipeline.

Before establishing the required test pressure, the designer must consider the following operating conditions of the system:

- Normal working pressure
- Maximum sustained operating pressure
- Maximum transient pressure along the pipeline

As an example, consider a water main installed in a system with the following operating conditions at the most critical location:

- Working pressure—60 psi
- Maximum sustained operating pressure—85 psi
- Maximum transient pressure—110 psi

In this case, a test pressure of 132 psi would provide a test pressure of 120 percent of the maximum transient pressure, 155 percent of the maximum operating pressure, and 220 percent of the working pressure. If the water main elevation is 202 ft and the elevation of a hydrant at the critical location is 208 ft, the required hydraulic grade line for the test is 507 ft (132 psi = [507 – 202] × 0.43). Therefore, the required gauge pressure as measured at the hydrant is 129 psi.

DURATION OF TESTS

The duration of the hydrostatic pressure test should be specified in the contract documents. Tests should be conducted over periods long enough to determine the average test allowance rates. The tests should be conducted over a day or more, although the minimum time duration should not be less than two hours.

TEST ALLOWANCE

The test allowance should be identified in the contract documents. In addition to measuring the makeup water added, all exposed portions of the system should be examined during the pressure test. All visible leaks should be repaired regardless of the test allowance.

It is recommended that no installation will be accepted if the quantity of makeup water added is greater than that determined by the following formula:

$$Q = \frac{LD\sqrt{P}}{148{,}000} \qquad \text{(Eq 8-1)}$$

where:

Q = quantity of makeup water, gph
L = length of pipe section being tested, ft
D = nominal diameter of the pipe, in.
P = test pressure, psi

This formula is based on a test allowance of 10.5 gallons/inch of nominal diameter/mile/day at a pressure of 150 psi. Table 8-1 summarizes makeup water allowances for a range of test pressures and diameters.

Do not test the water main in sections that are too long. With long test sections, large leaks may go undetected because of the averaging effect. The designer must consider placement of isolation valves with testing in mind to accommodate reasonable test sections.

WARNING: *There is serious safety hazard potential associated with testing components and systems with compressed air or other compressed gases.* The hydrostatic testing described in this chapter should only be conducted with water or other environmentally safe, incompressible fluids.

Table 8-1 Makeup water test allowances per 1,000 ft (305 m) of PVC pipe,* gph†

Avg. Test Pressure		Nominal Pipe Diameter, in. (mm)															
		4	6	8	10	12	14	16	18	20	24	30	36	42	48	54	60
psi	*(kPa)*	(100)	(150)	(200)	(250)	(300)	(350)	(400)	(450)	(500)	(600)	(750)	(900)	(1,050)	(1,200)	(1,350)	(1,500)
300	(2,065)	0.47	0.70	0.94	1.17	1.40	1.64	1.87	2.11	2.34	2.81	3.51	4.21	4.92	5.62	6.32	7.02
275	(1,895)	0.45	0.67	0.90	1.12	1.34	1.57	1.79	2.02	2.24	2.69	3.36	4.03	4.71	5.38	6.05	6.72
250	(1,725)	0.43	0.64	0.85	1.07	1.28	1.50	1.71	1.92	2.14	2.56	3.21	3.85	4.49	5.13	5.77	6.41
225	(1,550)	0.41	0.61	0.81	1.01	1.22	1.42	1.62	1.82	2.03	2.43	3.04	3.65	4.26	4.86	5.47	6.08
200	(1,380)	0.38	0.57	0.76	0.96	1.15	1.34	1.53	1.72	1.91	2.29	2.87	3.44	4.01	4.59	5.16	5.73
175	(1,205)	0.36	0.54	0.72	0.89	1.07	1.25	1.43	1.61	1.79	2.15	2.68	3.22	3.75	4.29	4.83	5.36
150	(1,035)	0.33	0.50	0.66	0.83	0.99	1.16	1.32	1.49	1.66	1.99	2.48	2.98	3.48	3.97	4.47	4.97
125	(860)	0.30	0.45	0.60	0.76	0.91	1.06	1.21	1.36	1.51	1.81	2.27	2.72	3.17	3.63	4.08	4.53
100	(690)	0.27	0.41	0.54	0.68	0.81	0.95	1.08	1.22	1.35	1.62	2.03	2.43	2.84	3.24	3.65	4.05
75	(515)	0.23	0.35	0.47	0.59	0.70	0.82	0.94	1.05	1.17	1.40	1.76	2.11	2.46	2.81	3.16	3.51
50	(345)	0.19	0.29	0.38	0.48	0.57	0.67	0.76	0.86	0.96	1.15	1.43	1.72	2.01	2.29	2.58	2.87

* If the pipeline under test contains sections of various diameters, the allowable makeup water allowance will be the sum of the computed test allowance for each size.

† To obtain makeup water allowance in liters per hour, multiply the values in the table by 3.79.

TEST ACCEPTANCE

A water main test should be considered acceptable if the gage pressure has not dropped by more than 5 psi and the testing allowance has not been exceeded when the test has been conducted for the duration specified in the contract documents or two hours, whichever is longer. In addition, there shall be no leaks. If a section fails to pass the hydrostatic field test, it will be necessary to locate, uncover, and repair or replace any defective pipe, valve, joint, or fitting. The pipeline must then be retested.

DISINFECTING WATER MAINS

Potable water mains should always be disinfected before being put into service. This very important procedure should be done under the direct control of the operating authority or its agents. Disinfection should comply with ANSI/AWWA C651.

The disinfection should take place after the initial cleaning and after the completion of the hydrostatic testing. Water from the existing water distribution system can be allowed to flow into the section of water main to be disinfected at a controlled rate.

All chlorinated water used for testing, flushing, or disinfecting water mains should be disposed of safely. High chlorine residuals may cause significant negative health effects on wildlife and humans. Confirm acceptable levels of chlorine with local, state, or federal authorities.

If the chlorine residual in the water main under disinfection remains relatively high, then usually there are no impurities in the systems creating a chlorine demand. However, if the chlorine applied to the section is gone, it can be concluded that there are impurities in the system.

After the water main has been recharged after successful chlorination, samples should be taken 24 hr after refilling for bacteriological tests according to ANSI/AWWA C651.

The system must not be put into service until the operating authority or its agents are satisfied with the testing procedure. No potable water pipeline should ever be put into service until the operating authority or its agents are satisfied with the testing procedure and post-disinfection sample test results.

In some trenchless installations like pipe bursting, fused PVC or mechanically joined PVC can also be pre-chlorinated before installation to expedite reinstatement of service and eliminate the need for bypassing. Check with the installer and/or pipe supplier for a detailed procedure.

SYSTEM MAINTENANCE

For PVC water mains, as well as those made from other materials, it is important to set up a system maintenance program. A good maintenance program accomplishes the following:

- helps prevent failure of facilities,
- detects problems in the water systems,
- determines the necessary replacement parts,
- provides input for future installation,
- maintains good public relations,
- detects and eliminates safety hazards,
- distributes maintenance work more advantageously, and
- reduces cost of system maintenance.

Records should be kept of all components of the work systems including date of installation, materials, size, pressure class, operating pressures, soil conditions, installer's name, and any other relevant information.

Records should be kept of pipe breakage, leak surveys, pressure tests, and friction coefficient tests to assist in decision making relating to maintenance and replacement of water distribution facilities.

The high dielectric strength of PVC pipe prevents the location of buried PVC pipe with electric current type metal detectors unless a tracer has been buried with the pipe during installation. Location during maintenance and future construction can be accomplished using any of a growing number of buried pipe detection devices and/or methods designed to locate nonmetallic buried pipelines. Probing with metal rods is not recommended. A summary of the methods for locating buried PVC pipelines is included in Chapter 12.

Frozen PVC water lines may be thawed by hot-water or steam injection. Torches and other direct-heating devices should not be used to thaw frozen lines.

REFERENCES

American National Standards Institute/American Water Works Association (ANSI/AWWA). 2013. C605. Underground Installation of PVC Pressure Pipe and Fittings for Water. Denver, Colo.: AWWA.

ANSI/AWWA. 2014. C651. Standard for Disinfecting Water Mains. Denver, Colo.: AWWA.

ASTM International. 2012. D2774. Standard Practice for Underground Installation of Thermoplastic Pressure Piping. West Conshohocken, Pa.: ASTM International.

Uni-Bell PVC Pipe Association. 2012 (5th ed.). *Uni-Bell Handbook of PVC Pipe—Design and Construction*. Dallas, Tex.: Uni-Bell PVC Pipe Association.

This page intentionally blank.

Chapter 9

Service Connections

This chapter provides recommendations for tools, procedures, and acceptable pipe products for adding branch lines to existing mains.

PIPE TAPPING OVERVIEW

In the pipe industry, "tapping" is the process of connecting a branch line to a main line by cutting a hole into the main. For tapping of PVC pressure pipe, branch lines can range in size from small service lines that supply individual homes to large outlets for industrial users. All taps—whether through service saddles, sleeves, or direct taps—should be perpendicular to the pipe. Service connections to PVC water mains are accomplished in the field via:

- direct tapping into the pipe wall,
- tapping through service saddles,
- tapping large service connections through tapping sleeves, or
- using fabricated or injection-molded couplings with threaded outlets. Guidance for outlet sizes is provided in Table 9-1.

Fabricated or injection-molded tapped couplings are in-line fittings that are installed during original construction. This type of connection is not included in this chapter; the coupling manufacturer should be consulted for information.

This chapter describes adding branch lines to existing mains using the following types of taps.

- Direct taps: The branch line is connected to the main by a corporation stop (Figure 9-1) that is screwed directly into the pipe wall. Maximum branch line size is 1 in.
- Saddle taps: The branch line is connected to a main by a corporation stop that is screwed into a metal saddle that wraps around the pipe (Figure 9-2). Maximum branch line size is 2 in.
- Sleeve taps: The branch line is joined to the main via a connection at a metal tapping sleeve that wraps around the pipe (Figure 9-3). Maximum branch size is the same as the main.

Table 9-1 Guidance for outlet sizes

				Tapping AWWA Pipe		
Pipe Material	Pipe Standard	Size (in.)	DR	Tapping Type: Direct Tap	Saddle Tap	Sleeve Tap
PVC	C900	4	All	*Not recommended*	≤2 in. diameter	Up to size-on-size
		6	25	*Not recommended*	≤2 in. diameter	Up to size-on-size
			18	≤1 in. diameter	≤2 in. diameter	Up to size-on-size
			14	≤1 in. diameter	≤2 in. diameter	Up to size-on-size
		8	25	*Not recommended*	≤2 in. diameter	Up to size-on-size
			18	≤1 in. diameter	≤2 in. diameter	Up to size-on-size
			14	≤1 in. diameter	≤2 in. diameter	Up to size-on-size
		10	All	≤1 in. diameter	≤2 in. diameter	Up to size-on-size
		12	All	≤1 in. diameter	≤2 in. diameter	Up to size-on-size
		14	≥41	*Not recommended*	≤2 in. diameter	Up to size-on-size
			≤32.5	≤1 in. diameter	≤2 in. diameter	Up to size-on-size
		16	≥51	*Not recommended*	≤2 in. diameter	Up to size-on-size
			≤41	≤1 in. diameter	≤2 in. diameter	Up to size-on-size
		≥18	All	*Not recommended*	≤2 in. diameter	Up to size-on-size
PVCO	C909	All	All	*Not recommended*	≤2 in. diameter	Up to size-on-size

DR: dimension ratio, PVC: polyvinyl chloride, PVCO: molecularly oriented polyvinyl chloride
Source: Adapted from Uni-Bell PVC Pipe Association
Notes: Direct tapping is not allowed in fused PVC pipe. The values in this table are recommendations only and the manufacturer should be consulted for guidance.

Source: The Ford Meter Box Company, Inc.

Figure 9-1 Corporation stop

Source: Uni-Bell PVC Pipe Association

Figure 9-2 Brass tapping saddle

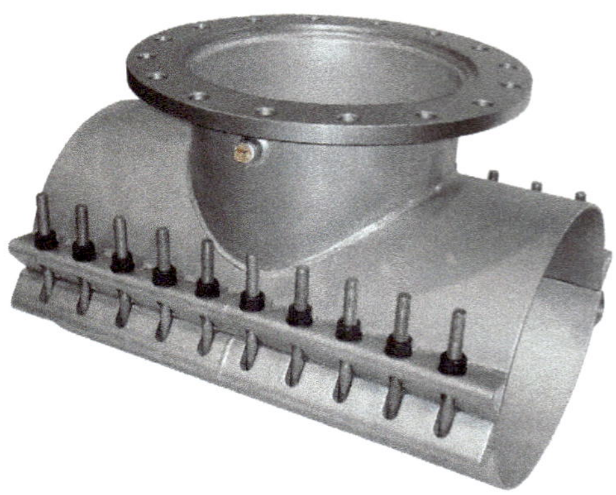

Source: The Ford Meter Box Company, Inc.

Figure 9-3 Tapping sleeve

Safety Considerations

During the drilling or tapping of any pressurized pipe, basic safety precautions should be taken to ensure personal safety of workers in the event of a sudden pipe failure. Although such situations are extremely infrequent, the following safety precautions are recommended.

- When a worker is drilling or tapping pipe under pressure, a second worker should be in the immediate vicinity.
- In addition to normal protective clothing, goggles or face shields should be worn.
- Ladders should be provided in the work area for quick and safe exit.
- A protective blanket with a hole at its center to permit installation and operation of the tapping and drilling machine should be provided to cover the exposed area of the pipe.
- The tapping crew should be familiar with the location of valves and their proper operation in case depressurization of the line is needed.
- Air should be removed from pipes before tapping. Failure to vent entrapped air can create a hazardous condition.

Coupon

The cylindrical piece of plastic removed during tapping is known as a "coupon" (see Figures 9-4 through 9-7).

The coupon cut from the pipe wall should be examined after removal from the cutter head. A smooth, clean cut indicates the feed rate was correct and the cutter was functioning properly.

Melted plastic or rough striations on the cut surface of the coupon indicate a dull cutter, too rapid a feed rate, or lack of lubrication of the cutting tool.

A raised edge, or "crown" on the edge of the coupon where the cutter first contacted the pipe wall, indicates that the feed rate was excessive. A dull cutter results in plastic deformation of the pipe wall, i.e., localized "crowning" at the edge of the tapping hole. Under this condition, the cutter cannot efficiently remove the material being cut. The cutter is forced away from the cut, creating the crown effect.

If the cutter is being forced through the pipe wall (excessive feed rate), a plug of material can be displaced from the inside surface of the pipe as the cutter approaches the interior pipe surface. This is often referred to as "punch-through." Punch-through can result in cracking of the pipe wall. Punch-through can also result from the use of a shell cutter with insufficient depth capacity for the wall thickness of pipe being tapped.

If striations, raised edges, or melted plastic are observed on a coupon, the tapping procedure and the condition of the tools should be re-examined and corrected before

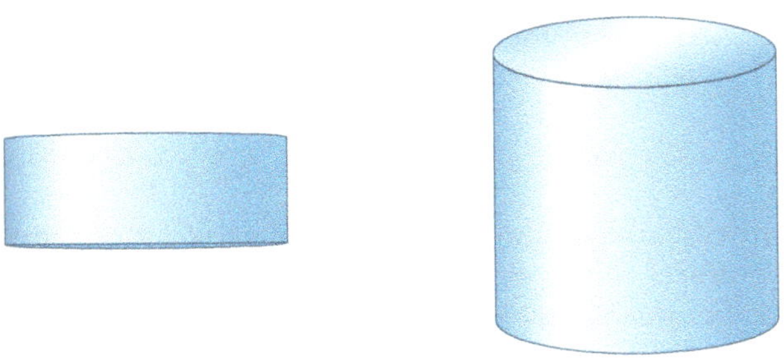

Source: Adapted from Uni-Bell PVC Pipe Association

Figure 9-4 Smooth coupon

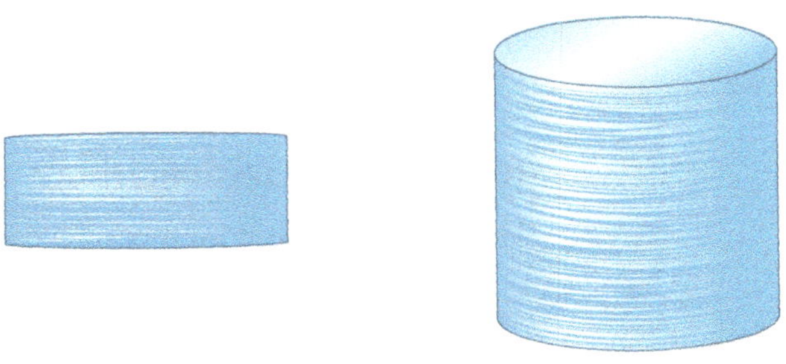

Source: Adapted from Uni-Bell PVC Pipe Association

Figure 9-5 Coupon with striations

Figure 9-6 Coupon showing "crown" at outer surface

Figure 9-7 Coupon showing "punch-through" at inner wall

additional taps are attempted. The engineer should be aware of the various thread types and make sure that the corporation stop threads match the threads in the pipe wall or saddle or sleeve. While the AWWA/taper/Mueller is the thread typically used for service lines, the IPS/NPT* thread type may be used when other items are connected to the PVC pipe.

Temperature Considerations

The equipment and methods recommended in this manual are also applicable for tapping in cold temperatures and in hot temperatures. No special considerations are necessary. Tapping can be accomplished at any temperature at which personnel and equipment can reasonably operate.

* Iron Pipe Size, National Pipe Thread

Literature

The following literature provides additional information and recommendations. Tapping crews should be familiar with

- AWWA Standard C605, "Underground Installation of Polyvinyl Chloride (PVC) and Molecularly Oriented Polyvinyl Chloride (PVCO) Pressure Pipe and Fittings" and
- AWWA Field Guide: "Pipe Tapping"

Notes:

1. This tapping chapter is not intended to take the place of the designer's judgment.
2. Illustrations used in this chapter depict specific tooling for the sake of clarity. Illustrations are not intended to promote specific types of equipment or the equipment of a particular manufacturer. Other equipment meeting the manual's requirements may be equally satisfactory.

DIRECT TAPPING

When a pipe is direct tapped, a hole is drilled into the pipe wall and threads are cut. The same machine that drills the hole is also used to cut the threads and to screw the corporation stop into the hole and tighten it securely.

The maximum outlet size for direct tapping is 1 in.

Table 9-2 provides guidance on which products are recommended for direct tapping.

Table 9-2 Guidance for direct tapping

			Direct Tapping AWWA Pipe	
Pipe Material	Pipe Standard	Size (in.)	DR	Comment
PVC	C900	4	All	*Not Recommended*
		6	25	*Not Recommended*
			18	≤1 in. diameter
			14	≤1 in. diameter
		8	25	*Not Recommended*
			18	≤1 in. diameter
			14	≤1 in. diameter
		10	All	≤1 in. diameter
		12	All	≤1 in. diameter
		14	≥41	*Not Recommended*
			≤32.5	≤1 in. diameter
		16	≥51	*Not Recommended*
			≤41	≤1 in. diameter
		≥18	All	*Not Recommended*
PVCO	C909	All	All	*Not Recommended*

Source: Adapted from Uni-Bell PVC Pipe Association

DR: dimension ratio, PVC: polyvinyl chloride, PVCO: molecularly oriented polyvinyl chloride

Notes: Direct tapping is typically not performed on 18-in. and larger PVC pipe. Direct tapping is not allowed on fused PVC pipe.

Equipment

Tapping machine. Several tapping machines are available to make direct service taps. The machines vary in design and operation depending on the manufacturer.

The machine chosen should provide the following features:

- Adapter base: The machine's adapter base should be properly sized to seal securely against the PVC pipe.
- Cutting tool: The machine should operate with a shell-design cutting tool suitable for PVC pipe.
- Hand rotation: The tapping machine should operate using a standard ratchet handle on the boring bar to rotate the boring bar by hand.
- Manual advance: The machine should be capable of manually advancing the boring bar independently from the rotation (i.e., no "automatic" advance). Typically, this is controlled by a feed nut and yoke or similar device.

When using a tapping machine, consider the following cautionary notes.

- Chains: Many direct tapping machines use a chain to secure the machine against the pipe. The use of a chain is not recommended for PVC pipe. If it is necessary to secure the machine using a chain when tapping PVC, the pipe should be adequately protected to prevent damage when the chain is tightened.
- Hand-held drill: A hand-held drill and brace and bit should never be used to tap PVC pipe (Figure 9-8).

Cutting/tapping tool. A tool used to direct-tap PVC is a specialized device such as a slotted core cutter (Figure 9-9) that bores a hole in the pipe wall and cuts threads to match those on the corporation stop that will be installed.

The cutting/threading tool used to direct-tap PVC pipe should:

- have a minimum of 1 slot for taps ⅝ in. or less;
- have a minimum of 2 slots for taps ¾ in. and larger;

Source: OlegSam/Shutterstock.com, warat42/Shutterstock.com

Figure 9-8 Cordless drill and brace and bit

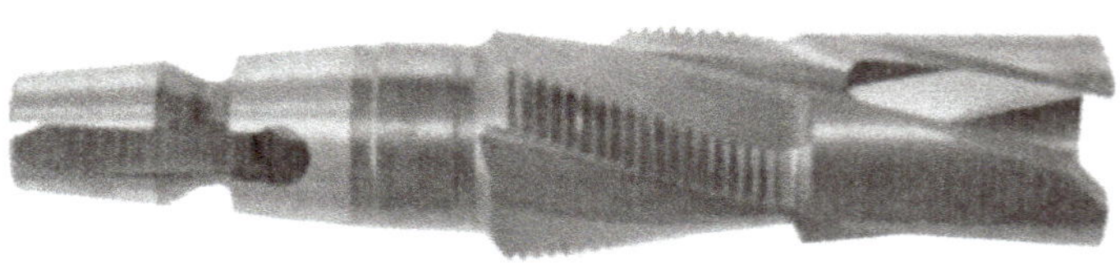

Figure 9-9 Slotted core cutter with threads for direct tapping

Source: Uni-Bell PVC Pipe Association

Figure 9-10 Improper tapping bits

- retain the coupon;
- provide a simple means of removing the retained coupon;
- have sufficient depth to accommodate the thickness of the wall being tapped;
- thread the hole with a thread pattern that matches the corporation stop that will be installed (typically AWWA C800 tapered threads);
- have a shank compatible with the cutting machine being used; and
- direct the shavings from the cut surface into the throat of the shell cutter.

When using a cutting/tapping tool, consider the following cautionary notes.

- Many shell cutters are designed to cut through smaller-diameter pipes with less wall thickness than found in larger-diameter PVC pipes. Consequently, some cutters do not have adequate throat depth to handle the heavier-walled pipe (e.g., AWWA C900, 12-in., DR 14 with average wall thickness 0.97 in.; C900, 16-in., DR 14 with average wall thickness 1.28 in.).
- Use of a twist drill, spade bit, auger bit, wood bit, hole saw, or similar cutting device is forbidden and may cause the pipe to crack (Figure 9-10).

Corporation stop ("corp stop"). Corp stops should be AWWA-tapered with thread complying with AWWA C800 in sizes ⅝ in., ¾ in., and 1 in. When a tap larger than 1 in. is required, the pipe should not be direct-tapped:

- For taps up to 2-in. diameter, see "Saddle Tapping"
- For taps greater than 2-in. diameter, see "Sleeve Tapping"

Procedures

Planning a direct tap. Following are the procedures for planning a direct tap.

- Outlet size: up to 1-in. maximum diameter
- Pipe product: See Table 9-2
- Wet/dry taps: Wet taps (pipe filled) and dry taps (pipe empty) are allowed. Wet taps can be made on pipe pressurized up to the full pressure class/rating of the pipe. However, the system operator may choose to lower the system pressure prior to tapping.
- Position of the tap
  - For 12-in. pipe or smaller, the tap should not be located closer than 24 in. from
    - The back of the bell (where the bell transitions to the barrel of the pipe)
    - The spigot insertion line
    - Joint-restraint hardware (Figure 9-11)

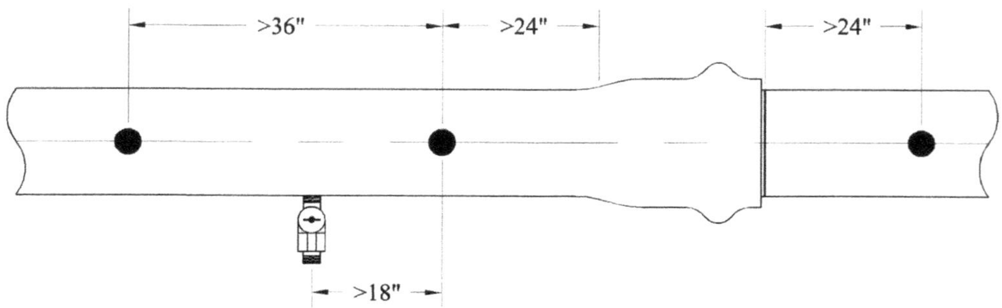

Source: Uni-Bell PVC Pipe Association

Figure 9-11 Tap spacing for 12-in. pipe or smaller

- For 14-in. pipe or larger, the tap should not be located closer than 36 in. from
  - The back of the bell (where the bell transitions to the barrel of the pipe)
  - The spigot insertion line
  - Joint-restraint hardware
- For all pipe, stagger multiple taps and keep them at least 18 in. apart lengthwise. Thus, the minimum spacing along the same line is 36 in.

Considerations for direct tapping. The following items must be considered when direct tapping.

- Safety: The tapping crew should be familiar with the precautions discussed in the Introduction section, "Safety Considerations."
- Reduced flow: Some operators choose to partially operate the control valves to reduce flow in the section being tapped. However, reducing the flow does not reduce system pressure.
- Manufacturer's instructions: For direct taps, the installation procedures may vary by machine manufacturer. It is important to comply with the manufacturer's recommended procedures when installing the required devices.
  - To minimize pipe distortion, the tapping machine should be installed on the pipe in accordance with the manufacturer's directions.

Adapted from Uni-Bell PVC Pipe Association

- Ease of cutting: PVC is relatively easy to cut when compared to metal. When PVC is properly tapped, the cutter turns easily with light pressure (Figure 9-12).
  - The force required to maintain a correct feed rate can be judged by "finger pull"—the effort required should be similar to opening a desk drawer.
  - The temptation to overfeed the cutter into the pipe wall should be resisted.
- Boring bar: When a pressurized pipe is tapped, system pressure will create an upward force on the boring bar. A 1-in. tap will result in about 0.8 lb of force per 1 psi of pressure in the system. This force will act to push the boring bar back toward the operator. It is very important to follow the machine manufacturer's instructions to safely control this effect.
- Reference marks: The hole must extend through the entire pipe wall. If the machine being used does not provide reference marks for tapping PVC, bench tests must be performed to develop these reference points and to mark the machine appropriately. "Cast iron" reference marks that may be provided by the machine manufacturer should not be used when tapping PVC pipes.

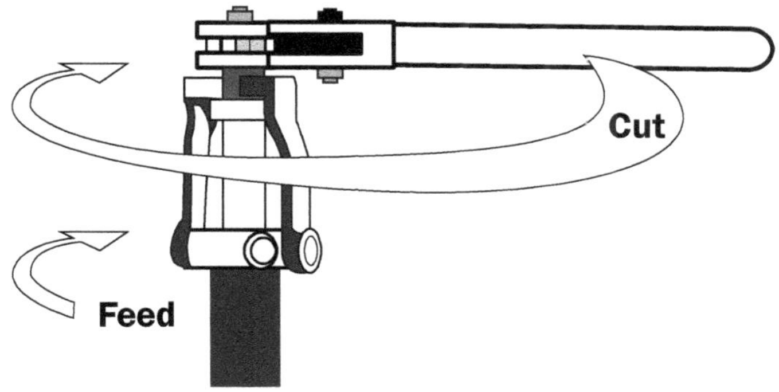

Source: Adapted from Uni-Bell PVC Pipe Association

Figure 9-12 Feed lightly, just enough to keep the cutter engaged

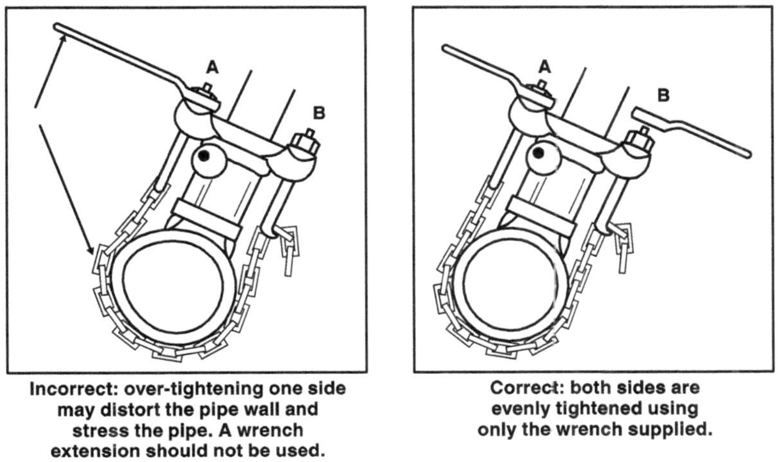

Source: Adapted from Uni-Bell PVC Pipe Association

Figure 9-13 Incorrect and correct mounting of tapping machine

Making a direct tap. Following are the procedures for making a direct tap.

- Control valves: Before tapping pressurized pipes, the crew should have a clear understanding of the valve operations necessary to isolate the tapping site and should ensure that the appropriate control valves are operational.
- Air: Before a pipe is tapped, air should be removed from the pipeline. Failure to vent entrapped air can create a hazardous condition.
- Mounting the machine
  - A correctly sized adapter should be mounted on the drilling machine. The adapter should match the pipe OD to provide a satisfactory seal.
  - The drilling machine should sit firmly on the pipe but should not distort the pipe (Figure 9-13).
  - The machine manufacturer's procedures for positioning and securing the drilling machine should be followed.
  - When direct tapping a pressurized line, an improperly matched chamber and cutter can block the valve, making it unusable. The operator should verify that the control valve can be closed by fully retracting the boring bar with the cutter

installed and closing the valve on the chamber after the drilling machine is installed.

- Cutting the hole and tapping the threads.
  - When making a direct tap in PVC pipe, the tapping machine manufacturer's operating procedures should be followed.
- The cutter should be lubricated per the manufacturer's instructions.
  - When PVC is properly tapped, the cutter turns easily with a light pressure. Feeding the cutter into the pipe wall too quickly should be avoided to prevent build-up of heat.
  - An accepted "rule of thumb" is to rotate the boring bar one complete turn for every one-eighth turn of the feed yoke. This allows the cutter to work as a cutter.
  - The force required to maintain a correct feed rate can be judged by "finger pull"—the effort required should be similar to opening a desk drawer.
  - When the finger pull criterion is applied in cold weather, the rate of feed will be reduced due to the properties of PVC.
  - The feed yoke is used to engage the tapping tool in the bored hole. After the threads are established in the bore, the tool becomes self-feeding and the feed yoke is typically disengaged from the boring bar. The machine manufacturer's operating instructions should be followed to complete the tap.
  - The pipe wall should be tapped to the correct depth. If the machine being used does not provide reference marks for tapping PVC, bench tests should be performed to develop these reference points and to mark the machine appropriately. "Cast iron" reference points provided by the machine manufacturer should not be used for tapping PVC pipes.
  - After the threads are properly completed, the tapping tool should be removed from the hole. To disengage the cutter and safely retract the boring bar, the machine manufacturer's instructions should be followed.
  - The coupon should be removed from the cutter and inspected per the Introduction section titled "Coupon."
- Consider the following cautionary notes when tapping pipe.
  - Wrench extenders ("cheater bars") should never be used when tapping PVC pipes.
  - When a pressurized pipe is tapped, system pressure will create an upward force on the boring bar. A 1-in. diameter tap will result in about 0.8 lbs of force per 1 psi of pressure in the system. This force will act to push the boring bar back toward the operator. To prevent injury, it is important to follow the machine manufacturer's instructions to safely control this effect.
- Following are instructions for inserting the corporation stop.
  - A direct tap is completed by using the drilling machine to install a corporation stop into the drilled and threaded hole that has been prepared in the pipe wall. Again, the machine manufacturer's operating procedures should be followed.
  - The threads of the stop should be properly lubricated by applying two clockwise spiral wraps of Teflon tape. Other thread lubricants are not recommended. Liquid sealants (even though they may contain Teflon) should not be used.
  - The corporation stop is attached to the boring bar. The stop should be in the closed position to prevent flow when the machine is removed after the installation is completed.

- The tapping machine is reattached to the tapping chamber, the valve is opened, and the stop is inserted into the prepared tap hole.
- The machine's operating procedures should be followed to tighten the corporation stop, disengage the boring bar, and remove the machine from the main.
- After the machine is removed, the tap should be inspected. At correct insertion depth, two to three threads should be visible on the corporation stop.
- If leakage is occurring around the threads, the stop can be tightened. Exceeding 35 ft-lb of torque to stop any leakage around the threads is not recommended.
- If leaking past the threads persists, it may be necessary to remove the pressure from the line, remove the corporation stop, and clean the threads. When reinstalling the stop, Teflon tape should be reapplied to the threads and the manufacturer's torque recommendations should be followed.
- When making a direct tap on PVC pipe that is empty, not in service, or not yet under pressure, some of the procedures can be modified. For example, if the machine manufacturer's operating instructions allow, the tapping machine can be removed after the tap is completed and the corporation stop can be installed by hand.

Safety Considerations

Before any tapping activities are performed, it is recommended that the tapping crew review the Introduction section titled "Safety Considerations."

SADDLE TAPPING

When a pipe is saddle-tapped, equipment is attached to the corporation stop that permits a cutting tool to be fed through the stop to cut a hole in the pipe. No threading of the pipe wall is required since the corporation stop is screwed into the service saddle.

The use of saddles to make taps is recommended for all sizes and classes of PVC and PVCO pressure pipe, including fused PVC pipe. Service connections up to 2-in. size may be made using a service saddle. For fused PVC pipe, the supplier should be consulted for the range of tap sizes for each pipe size. Table 9-3 provides guidance for saddle tapping.

Another term for "tapping saddle" is "service saddle."

Equipment

Tapping saddles. When a tapping saddle is used, the tapping machine attaches to a corporation stop that is screwed into the saddle outlet. This allows a cutting tool to be

Table 9-3 Guidance for saddle tapping

		Saddle Tapping AWWA Pipe		
Pipe Material	Pipe Standard	Size (in.)	DR	Comment
PVC	C900	All	All	≤2 in. diameter
PVCO	C909	All	All	≤2 in. diameter

Source: Adapted from Uni-Bell PVC Pipe Association

Note: In the "Saddle Tapping" section, "PVC" applies to both PVC (including fused PVC) and PVCO pipe.

fed through the corporation stop to cut a hole in the pipe. No threading of the pipe wall is required as the corporation stop is threaded into the saddle.

- Many saddles are manufactured to accommodate multiple piping materials. As a result, despite the stated nominal diameter, the typical dimensional tolerances are very broad. Since the outside diameter of PVC pipe is controlled to strict tolerances, these "wide-range" saddles should not be used for PVC pipe.
- The saddle chosen should
  - Be designed and sized for use on PVC pipe (see Figures 9-14 and 9-15 for improper and proper saddle sizing, respectively)
  - Provide full support around the circumference of the pipe
  - Provide a bearing area of sufficient width along the longitudinal axis of the pipe to minimize pipe distortion when the saddle is properly tightened:
    - For taps up to 1 in. diameter, a minimum of 2 in. total width along the pipe axis
    - For taps from 1¼ in. through 2 in., a minimum of 3 in. total width
    - Minimum width of any individual band not less than 1 in.

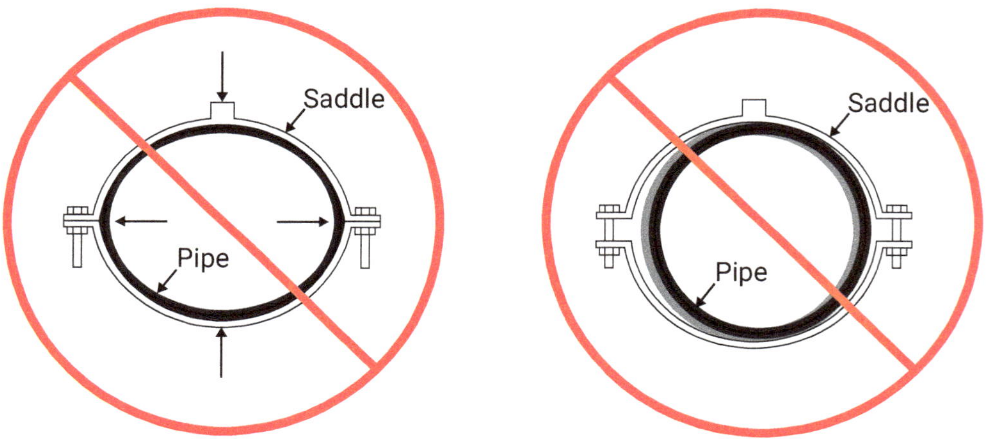

Source: Adapted from Uni-Bell PVC Pipe Association

Figure 9-14 Improper saddle sizing

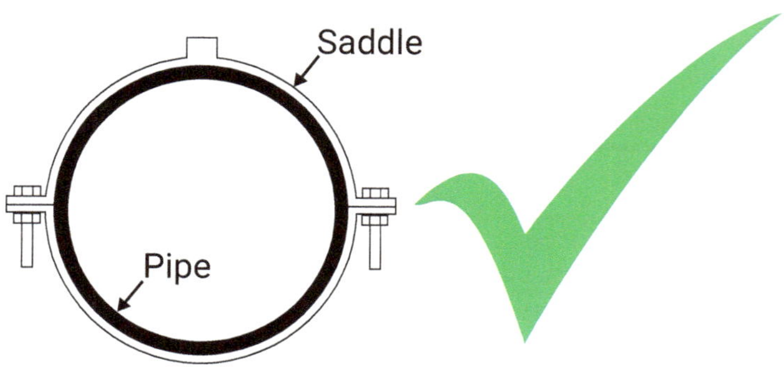

Source: Adapted from Uni-Bell PVC Pipe Association

Figure 9-15 Proper saddle sizing

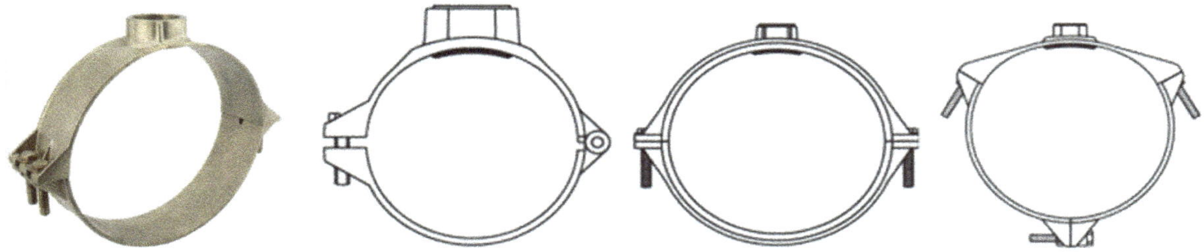

Source: Uni-Bell PVC Pipe Association

Figure 9-16 Brass tapping saddles

- Service saddles should not
  - Have lugs that will dig into the pipe when the saddle is tightened
  - Have U-bolt type straps that do not provide sufficient bearing area
- The service saddle manufacturer should be consulted for recommended pressure capacity.

Brass tapping saddles are depicted in Figure 9-16.

- **Corporation stop ("corp stop")**. The threads on the corp stop must match the threads on the outlet of the saddle.
- The maximum size of corp stop used with a service saddle is 2 in. When a tap larger than 2 in. is required, the pipe should not be saddle-tapped:
  - For taps greater than 2-in. diameter, see "Sleeve Tapping."

Tapping machine. Several tapping machines are available that will cut through a corp stop. Tapping machines vary in design and operation depending on the specific machine manufacturer.

- Tapping machines should
  - Use a shell-design cutting tool suitable for PVC pipe.
  - Operate using a standard ratchet handle on the boring bar to rotate the boring bar by hand.
  - Be capable of manually advancing the boring bar independently from the rotation (i.e., no "automatic" advance). Typically, this is controlled by a feed nut and yoke or similar device (Figure 9-17).
- Consider the following cautionary notes:
  - Hand-held drill: A hand-held drill and threading boring bar without ratchet feed (Figure 9-18) should never be used to tap PVC pipe.

Cutting tool. The cutting tool used to saddle-tap PVC pipe should

- Have a minimum of 1 slot for taps ⅝" or less
- Have a minimum of 2 slots for taps ¾" and larger
- Retain the coupon
- Provide a simple means of removing the retained coupon
- Have sufficient depth to accommodate the thickness of the wall being tapped
- Have a shank compatible with the cutting machine being used
- Direct the shavings from the cut surface into the throat of the shell cutter

Source: Uni-Bell PVC Pipe Association

Figure 9-17 Tapping machines with feed nut and yoke

Source: OlegSam/Shutterstock.com; Uni-Bell PVC Pipe Association

Figure 9-18 Hand-held drill and threading boring bar without ratchet feed

Source: Uni-Bell PVC Pipe Association

Figure 9-19 Improper cutters

- Consider the following cautionary notes when using cutting tools:
  - Many shell cutters are designed to through smaller-diameter pipes with less wall thickness than found in larger-diameter PVC pipes. Consequently, some cutters do not have sufficient throat depth to handle the heavier-walled pipe (e.g., AWWA C900 12-in. DR 14 with average wall thickness 0.97 in.; C900 16-in. DR 14 with average wall thickness 1.28 in.).
  - Use of a twist drill, spade bit, auger bit, wood bit, hole saw, or similar cutting device is not recommended and may cause the pipe to crack. Figure 9-19 shows examples of improper cutters.

Procedures

Planning a saddle tap. Following are the procedures for planning a saddle tap.

- Tapping hole size: Taps up to 2-in. diameter can be made through a service saddle.
- Wet/dry taps: Wet taps (pipe filled) or dry taps (pipe empty) can be made. Wet taps can be made on pipe pressurized up to the full pressure class/rating of the pipe.
- Position of the tap
    - For 12-in. pipe or smaller, the tap should not be located closer than 24 in. from
        - The back of the bell (where the bell transitions to the barrel of the pipe)
        - The spigot insertion line
        - Joint-restraint hardware
    - For 14-in. pipe or larger, the tap should not be located closer than 36 in. from
        - The back of the bell (where the bell transitions to the barrel of the pipe)
        - The spigot insertion line
        - Joint-restraint hardware
    - For all pipe, stagger multiple taps and keep them at least 18 in. apart lengthwise. Thus, the minimum spacing along the same line is 36 in.

Considerations for saddle tapping. The following items must be considered when saddle tapping.

- Safety: The tapping crew should be familiar with the precautions discussed in the Introduction section titled "Safety Considerations."
- Reduced flow: Some operators choose to partially operate the control valves to reduce flow in the section being tapped. However, reducing the flow does not reduce system pressure.
- Manufacturer's instructions: For saddle taps, the installation procedures may vary by manufacturer. It is important to comply with the manufacturer's recommended procedures when installing the required devices.
    - To minimize pipe distortion, the tapping saddles should be assembled on the pipe in accordance with the manufacturer's directions.
    - Some distortion may occur in thinner-walled PVC and PVCO pipes. This condition is acceptable as long as the saddle manufacturer has qualified the saddle by testing with the pipe product used.
    - Ease of cutting: PVC is relatively easy to cut when compared to metal. When PVC is properly tapped, the cutter turns easily with light pressure.
        - The force required to maintain a correct feed rate can be judged by "finger pull"—the effort required should be similar to opening a desk drawer.
        - The temptation to overfeed the cutter into the pipe wall should be resisted.
- Boring bar: When a pressurized pipe is tapped, system pressure will create an upward force on the boring bar. A 1-in. tap will result in about 0.8 lb of force per 1 psi of pressure in the system. This force will act to push the boring bar back toward the operator. It is important to follow the machine manufacturer's instructions to safely control this effect.
- Reference marks: The hole must extend through the entire pipe wall. If the machine being used does not provide reference marks for tapping PVC, bench tests must be performed to develop these reference points and to mark the machine

appropriately. "Cast iron" reference marks that may be provided by the machine manufacturer should not be used when tapping PVC pipes.

Making a saddle tap. The saddle should be installed on the pipe following the manufacturer's instructions.

- Recommended torque values should not be exceeded.
- The inlet side of the corporation stop is then screwed into the saddle threads. The stop hardware should be retained.
- The corporation stop should be opened.
- The cutter should be lubricated per the manufacturer's instructions.
- Using the correct adapter and gasket, the tapping machine is attached to the corporation stop's outlet threads.
- Before the tap is begun, the boring bar and cutter should be fully retracted, and the corporation stop opened and reclosed to ensure that there is no interference.
- Use a machine with an operator-controlled feed rate. The use of a shell cutter is essential. Follow the machine manufacturer's instructions.
- The operator should lower the boring bar to the main and rotate the cutter while exerting finger-pull on the feed handle. The ratchet handle should be rotated one complete turn for every one-eighth turn of the feed yoke to allow the cutter to work as a cutter.
- When the tap is completed, the cutter should be retracted, the corp stop closed, and the tapping machine removed.
- When retracting the boring bar and cutter from the tapped pipe, the machine manufacturer's operating instructions should be followed. Failure to do so may create a situation that endangers the personnel making the tap and may result in serious injuries.

Safety Considerations

Before any tapping activities are performed, it is recommended that the tapping crew review the Introduction section titled "Safety Considerations."

SLEEVE TAPPING

Large-diameter taps are greater than 2-in. nominal diameter in size. Tapping sleeves are recommended for these large-diameter connections. The use of sleeves to make taps is recommended for all sizes and classes of PVC and PVCO pressure pipe, including fused PVC pipe. For fused PVC, the supplier should be consulted for the range of tap sizes for each pipe size. Guidance for sleeve tapping is provided in Table 9-4.

Table 9-4 Guidance for sleeve tapping

		Sleeve Tapping AWWA Pipe		
Pipe Material	Pipe Standard	Pipe Size	DR	Comment
PVC	C900	All	All	Up to size-on-size
PVCO	C909	All	All	Up to size-on-size

Source: Adapted from Uni-Bell PVC Pipe Association

The lateral or branch being connected may be any diameter up to the diameter of the pipe main (size-on-size connection). When making a sleeve tap, the drilling equipment attaches to a valve that allows a cutting tool to be fed through the valve to cut a hole in the pipe. The actual diameter of the hole to be cut is less than the inside diameter of the pipe used for the branch and should never exceed the inside diameter of the main being tapped.

Note: In this chapter, "PVC" applies to both polyvinyl chloride (PVC) pipe (including fused PVC) and oriented polyvinyl chloride (PVCO) pipe.

Equipment

Tapping sleeve. The following information should be provided when specifying tapping sleeves (Figure 9-20).

- Outside diameter of the pipe being tapped
- Dimension ratio (DR) of pipe being tapped
- Size of the outlet desired
- System working pressure

Tapping sleeves should

- Provide full support around the circumference of the pipe
- Provide sufficient length to
  - Enable satisfactory sealing
  - Prevent potential movements of the pipe
- Be independently supported to minimize loading on the pipe being tapped
- Be compatible with the tapping valves and/or attaching mechanisms of the drilling machine that will be used

Tapping sleeves (Figure 9-20) should not

- Distort the pipe when properly tightened
- Have lugs or other protrusions that will dig into the pipe when the sleeve is tightened

Tapping valve. A tapping valve is used to control flow into the newly installed lateral line after the tap is completed. Tapping valves are typically gate valves.

- The valve opening must be of sufficient diameter to allow the cutter to pass through the valve to complete the tap.

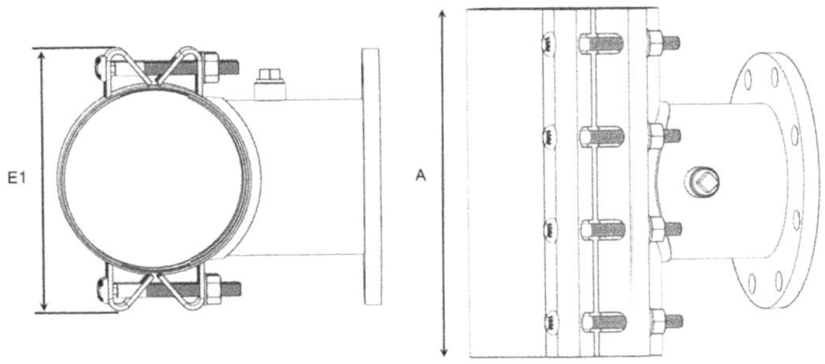

Source: The Ford Meter Box Company, Inc.

Figure 9-20 Tapping sleeve

- When tapping "dry" pipes, it is possible to tap without using a tapping valve by selecting an appropriate adapter for the drilling machine. However, this would not be a typical installation.
- A tapping valve should:
  - Match the connectors on both the tapping sleeve and the drilling machine
  - Allow the cutter to pass through without interference
  - Comply with the system operator's criteria if it will be used as a permanent control valve

Tapping machine. Several manufacturers produce tapping machines designed to make large-diameter taps. These machines vary in design and operation depending on the specific machine. For all taps, the machine and the valve used must mate properly.

- Most large-diameter tapping machines use power units that operate at very low rotation speeds. The tapping crew should ensure that the manufacturer of the machine being used recommends it for tapping PVC pipe. The crew should also follow the machine manufacturer's instructions.
- Large-diameter tapping equipment is available for rental or purchase from many vendors. In addition, in many areas, contractors who specialize in this type of work are available. Due to the expertise required to perform large-diameter taps, the use of experienced personnel is recommended.

Cutting tool. Because of the wide ranges of pipe size and pipe thickness encountered, several different types of cutting tools are used.

- The cutting tool should:
  - Have the capability to retain the coupon
  - Have sufficient depth to accommodate the thickness of the wall being drilled
  - Have a shank compatible with the cutting machine being used

Procedures

Planning a tap through a sleeve and valve

- Tapping hole size
  - Connections up to size-on-size can be made through a tapping sleeve and valve.
  - To prevent damage to the main, the diameter of the hole being cut should never exceed the inside diameter of the main.
  - Several tapping-sleeve manufacturers recommend that the hole cut in the main line be smaller than size-on-size. The recommendations of the sleeve manufacturer regarding hole size should be followed.
- Wet/dry taps: Both wet taps (pipe filled) and dry taps (pipe empty) can be made.
- Non-pressurized pipe: When tapping non-pressurized pipes, a valve may not be required.
- Position of the tap
  - For 12-in. pipe or smaller, the tapping sleeve should not be located closer than 24 in. from:
    - The back of the bell (where the bell transitions to the barrel of the pipe)
    - The spigot insertion line
    - Joint-restraint hardware

- For 14-in. pipe or larger, the tapping sleeve should not be located closer than 36 in. from:
  - The back of the bell (where the bell transitions to the barrel of the pipe)
  - The spigot insertion line
  - Joint-restraint hardware
- For spacing criteria for multiple sleeve taps, the sleeve manufacturer should be consulted.

Considerations for large-diameter tapping

- Safety: The tapping crew should be familiar with the precautions discussed in "Safety Considerations."
- Reduced flow: Some operators choose to partially operate the control valves to reduce flow in the section being tapped. However, reducing the flow does not reduce system pressure.
- Manufacturer's instructions: For large-diameter taps, the installation and operating procedures of the items required may vary by manufacturer. It is important to comply with the manufacturer's recommended procedures when installing and operating the required devices.
  - To ensure that the gasket will seal properly, the pipe should be cleaned, and the gasket should be lubricated and assembled in accordance with the sleeve manufacturer's directions.
  - To minimize pipe distortion, the tapping sleeve should be assembled on the pipe in accordance with the manufacturer's directions.
  - Some distortion may occur in thinner-walled PVC and PVCO pipes. This condition is acceptable as long as the sleeve manufacturer has qualified the sleeve by testing with the pipe product used.
- Structural support: Tapping sleeves should be well-supported independently from the pipe during tapping. Support should be left in place after tapping. It is important to realize that tapping sleeves are designed for sealing purposes only, not for structural support.
- Joint restraint: Thrust blocks or joint-restraint devices should be used as with any other fitting or appurtenance. If the tapping valve will be used as a permanent control valve, adequate blocking designed to accommodate the torque associated with valve operations should be included.
- Alignment of cutting tool: The larger the tapping hole size, the more important it is to ensure that the cutting tool is perpendicular to the axis of the pipe. This will prevent stresses caused by misaligned penetration of the ID.

Making a large-diameter tap

- The tapping sleeve should be assembled on the pipe in accordance with the manufacturer's directions, ensuring that minimal pipe distortion occurs.
- The tapping valve is then connected to the sleeve.
- Support blocks are positioned under the valve.
- The tapping machine and adaptor are then attached to the valve outlet flange.
- The tapping valve is opened, the cutter advanced, and the hole cut into the main through the sleeve.

- After the hole is cut, the cutter is retracted from the tapped pipe. The machine manufacturer's operating instructions should be followed. Failure to do so will create a dangerous situation that may result in serious injuries.
- The tapping valve is then closed, and the tapping machine removed.
- The coupon should be removed from the cutter and inspected per the Introduction section titled "Coupon."

Safety considerations. Before any tapping activities are performed, it is recommended that the tapping crew review the Introduction section titled "Safety Considerations."

REFERENCES

Uni-Bell PVC Pipe Association. 2012 (5th ed.). *Uni-Bell Handbook of PVC Pipe—Design and Construction*. Dallas, Tex.: Uni-Bell PVC Pipe Association.

Water Research Foundation. 2007. Impact of Petroleum-Based Hydrocarbons on PE/PVC pipes and Pipe Gaskets. Report 2946. Denver, Colo.: Water Research Foundation.

Chapter 10

Trenchless Installation of PVC Pipe

As our water and wastewater infrastructure ages, there is constant need for repair and replacement and as cities grow, there is a continual need for additional underground pipelines in already densely populated, complex environments. There are also situations that make traditional excavation costly and difficult (such as that presented by rivers or major roadway and wetlands that must be crossed). In all these applications, trenchless (or "no-dig") pipe installation methods are being used effectively. This chapter covers recommended and common practices for trenchless installation of PVC pipe.

Construction methods for trenchless installation have progressed to computer-operated directional-drilling rigs capable of pulling in thousands of feet of assembled PVC pipe (fused or segmented) in one pull, as well as slip-lining and pipe-bursting of pipe in disrepair. Meanwhile, PVC materials, pipe, and specialized joints have been developed to accommodate various trenchless installation methods.

Because unique loads are placed on trenchless pipe during installation, special attention must be given to the axial compressive and/or tensile forces on the pipe. These forces arise from pulling and pushing during installation and from the external loads on the pipeline once it is fixed in place.

PVC PROPERTIES FOR TRENCHLESS CONSTRUCTION

PVC plastic pipe possesses properties that are conducive for trenchless installations. Table 10-1 summarizes these and their importance.

Table 10-1 Important material properties for trenchless installation

Material Property	Value	Importance
Tensile strength	7,000 psi	Trenchless installation requires axial strength to allow restrained joined pipe to be pulled into place
Modulus of elasticity	400,000 psi	Key stiffness property for external load resistance
Coefficient of thermal expansion	3.0×10^{-5} in./in./°F	Low expansion and contraction, compared to the other common thermoplastic pipe material, minimizes axial forces generated by temperature changes
Hardness	≈ 117 (Rockwell R scale)	High resistance to abrasion and scratching during trenchless installation
Elasticity	No value	While PVC is classified as a thermoplastic and therefore visco-elastic in behavior, it behaves elastically over nearly all its stress–strain range, resulting in no wait period for reversion after tensile pull-in
Hydrostatic design basis (HDB)	4,000 psi	Greater long-term strength, higher HDB, than the other common pipe thermoplastic, which allows for thinner pipe walls, greater flow capacity, and less weight for the same pressure capacity/strength
Specific gravity	1.4	Having a density greater than water, PVC pipe will not float when ballasted with water
Allowable strain capacity for bending	Recommended minimum bend radii are provided in Chapter 12	Less strain capacity than the other common thermoplastic pipe material but more strain capacity than all other commonly used pipe materials and enough to accommodate the curvature requirements generally associated with trenchless installation

TRENCHLESS INSTALLATION JOINTS AVAILABLE WITH PVC

There are several PVC pipe joint options that can be used for trenchless installation. The strength and size of the joint will vary as well as the suitability for each trenchless method discussed.

For installation by horizontal directional drilling (HDD), the outside diameter of the joint (ODJ) will determine the size of the bore hole. Typical guidance is for the bore hole to be at least 1.5 times the ODJ up to 4-in. pipe and then ODJ plus 12 in. for larger sizes. Tensile and compression load capacity of each joint type will vary. Check with the pipe manufacturer/supplier for maximum ODJ and recommended load limits for each joint type. The manufacturer/supplier should also indicate and confirm what safety factor has been used in determining the recommended tensile load limit for both the pipe and joint.

Trenchless Joint Types

Butt fused joints. Plain end lengths of PVC pipe are joined by the thermal butt fusion process (Figure 10-1). The result is a continuous pipe that has practically no external profile when the fusion bead is removed.

Spline and grooves joints. A groove is cut into the outside of the pipe near the spigot end and also inside either the bell or coupling. When the inside and outside grooves are aligned, a plastic spline is inserted to restrain the connection (Figure 10-2). A gasket provides the watertight seal.

Pins and groove joints. The bell and spigot of the PVC pipe are thickened during extrusion and a groove is machined into the spigot with evenly spaced holes drilled into the bell. The spigot is then inserted into the bell to align the groove with the bell holes.

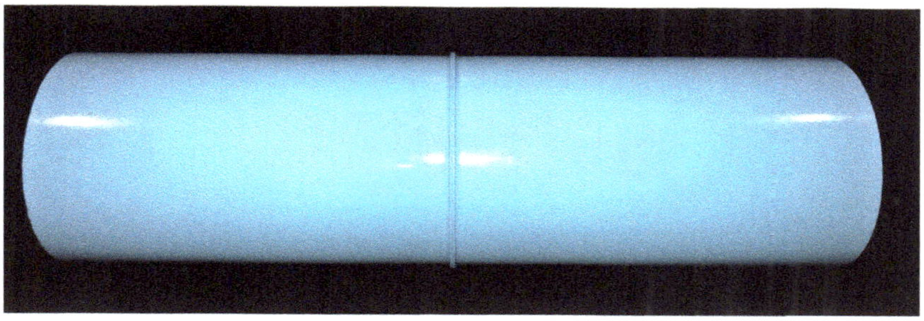

Source: Underground Solutions, Inc.

Figure 10-1 Butt fused pipe

Source: NAPCO Pipe & Fittings

Figure 10-2 Spline and grooves restraint

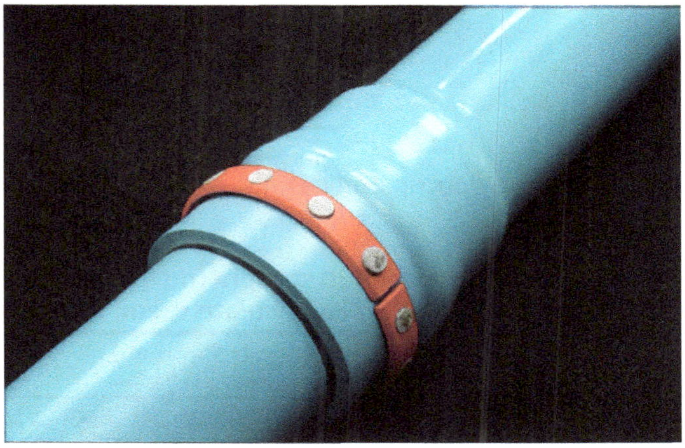

Source: Underground Solutions, Inc.

Figure 10-3 Pins and groove restraint

Fiberglass pins are then tapped through the bell into the spigot groove. A gasket makes the watertight seal (Figure 10-3).

Grip ring and casing joints. The gasket race-way in the pipe bell is extended to accommodate a tapered metal casing and metal grip-ring. The metal casing inside the pipe bell forces the grip ring into the pipe spigot when a pull-force is applied. A gasket makes the watertight seal (Figure 10-4).

Allowable Pull and Push Forces for PVC Pipe Joints

The anticipated maximum pull load during a trenchless pull-in should not exceed the allowable pipe joint pull force. The most severe pull loads are generally associated with the HDD installation method. The value is derived from the tensile properties of the pipe, as in the case of fusion-joined PVC, or from the strength of the mechanical connection, such as the pins and groove, spline and grooves, grip ring and casing, or externally harnessed restrained joint. An example is shown in Table 10-2. The provider of each joint type should be contacted to obtain their recommended allowable pull and push forces for their available joint sizes and dimension ratios (DRs). The recommended values should be evaluated to ensure that they exceed the expected maximum pull or push force needed for a specific project.

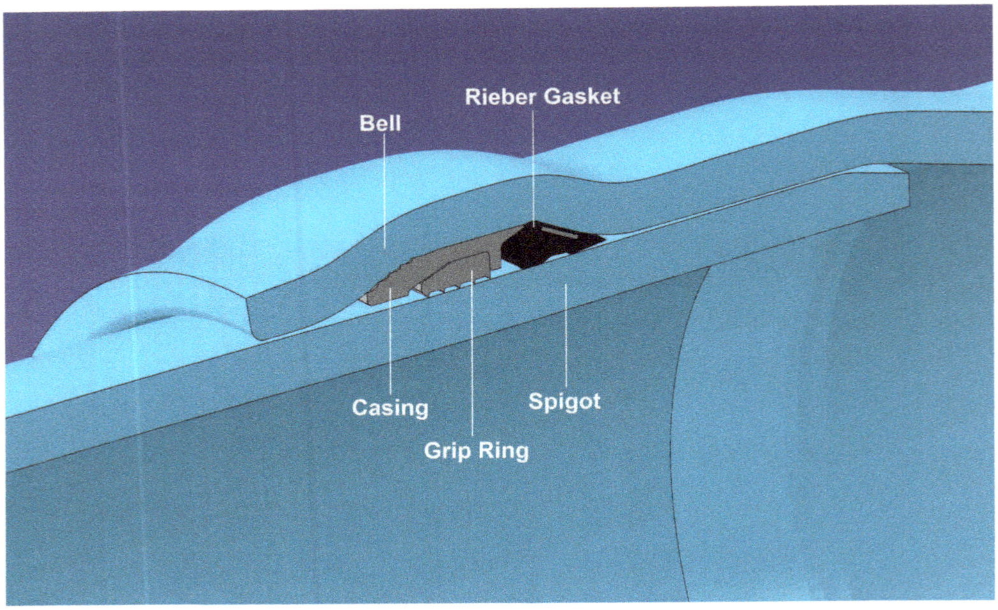

Source: S&B Technical Products

Figure 10-4 Grip ring and casing

Table 10-2 Recommended straight (no bending) pull and push force values, along with the safety factors used to calculate them, for 8-in. DR 18 PVC pipe

Joint Type	Allowable Tensile (Pull) Force, lb	Allowable Compression (Push) Force, lb	Applied Safety Factor
Fused	37,800	37,800	2.5
Spline-locked	27,500	27,500	2.0
Pinned	25,800	25,800	2.0
Grip-ring	30,000	N/A*	1.6

* Compressive forces are not applicable and/or applied to this joint.

In addition to the pipe properties noted in Table 10-1, other assumptions or properties must be considered to determine the maximum installation pull and push forces. For HDD installations, this will include a friction factor between the PVC pipe and drill mud (0.3 is commonly used), a friction factor between the PVC pipe and either the ground (0.5 is a typical value) or on rollers (values of 0.1–0.2 are typical). Soil densities and drill mud densities are also properties that impact the HDD pull force calculation. The general layout of the bore alignment including insertion/exit angles, radii of curvature, and depth may also need to be considered.

TRENCHLESS CONSTRUCTION AND TRENCHLESS REHABILITATION

Trenchless installation can be classified into two groups:

1. *Trenchless construction*: The installation of an entirely new pipeline with minimal open-cut excavation. Methods for new construction include
   a. Horizontal directional drilling (HDD)
   b. Installations through casing (jack and bore)

2. *Trenchless rehabilitation*: The repair of an existing deteriorated pipeline with minimal open-cut excavation. The primary rehabilitation methods are
   a. Sliplining
   b. Pipe bursting
   c. Tight-fit structural lining

Trenchless Construction—Horizontal Directional Drilling

HDD is the most commonly used trenchless process for installing new pipelines. HDD is performed with a drilling rig and involves three steps.

1. A pilot hole is bored.
2. The hole is reamed to its final size.
3. The pipe is pulled in.

In smaller diameters (<12 in.) and shorter length installations (<500 ft), the second and third steps are sometimes performed simultaneously. The design considerations of typical HDD alignment are depicted in Figure 10-5.

Bending. HDD bore alignments require some amount of bending of pipe and/or joints. Each PVC pipe joint technology provider publishes guidance on allowable minimum bend radii for pipe with its particular joining system (Figure 10-6). This guidance must be followed in the design of bore alignment as well as during pipe layout and installation.

The recommended degrees of allowable angular deflection at bell-and-spigot joints and couplers are dependent upon the joint geometry and gasket design and generally range from zero to five degrees. The recommended minimum bending radii for fusion-joined PVC pipe are included in Chapter 12, Table 12-2.

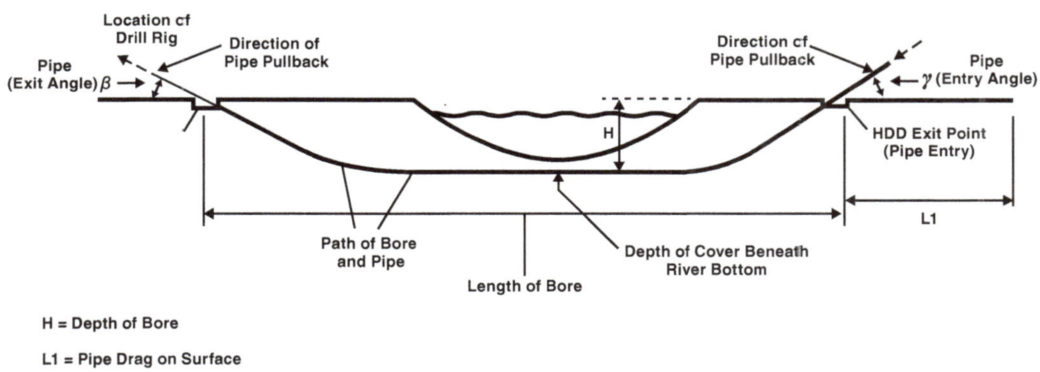

Figure 10-5 Typical HDD alignment shown schematically for design considerations

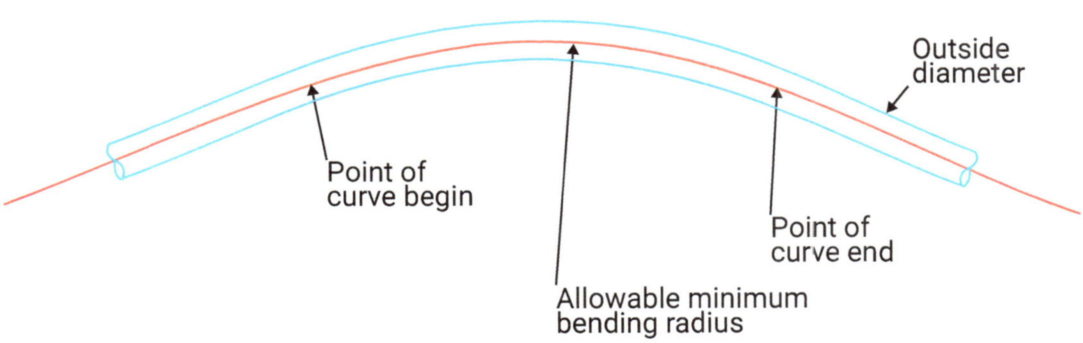

Source: Adapted from Underground Solutions, Inc.

Figure 10-6 Illustration of minimum bend radius

Pullback force. The combined effects of all of the frictional and resistive forces on the pipe create the pullback force required for successful pipe installation. The resultant pullback force required must not exceed the tensile capabilities of the pipe and/or joint together with an appropriate safety factor. Monitoring and controlling the density of the drilling fluid is critical for minimizing the required pull force.

Information on how to estimate pullback forces and allowable tensile stresses may be obtained by contacting the manufacturer or technology provider of the particular PVC pipe and restrained joining system selected for use.

Elongation and connections. With PVC pipe, the magnitude and duration of pulls do not result in significant pipe elongation. Therefore, connections to PVC pipe can be made immediately following pullback. Joint restraint is recommended when joining a directional drilled portion of a pipeline to another section of pipe, a fitting, or other appurtenance.

Pressure testing. Installers pressure testing HDD lines should be aware that the combination of drilling fluid and a reamed hole can allow the pipe to move and therefore all connection points should be adequately restrained for any thrust prior to the testing. For trenchless installations, the pressure test in ground after installation is the only test that occurs after all loading of the pipe is realized. Above-grade hydrotesting is not recommended.

Methods to reduce friction and minimize exterior scratching. Moving long lengths of pipe on roller stands, timbers, or short pipe sections or dragging pipe on grass or soft soil will help prevent potential damage to the pipeline and reduce the force needed to install the pipe.

Supporting the pipeline on rollers reduces the frictional drag coefficient and the pull force required to complete an installation (Figures 10-7 and 10-8). Steep insertion angles require additional support consideration so that the manufacturer's recommended angles and bend radii are not exceeded at the installation point.

The generally recommended minimum lengths between roller supports are listed in Table 10-3. The pipe supplier should be contacted for specific minimum spacing

Table 10-3 Pre-installation, recommended support spacing for restrained-joint PVC pipe (empty)*

Nominal Pipe Diameter		Max. Support Spacing†	
(in.)	*(mm)*	*(ft)*	*(m)*
4	100	13	4.0
6	150	17	5.2
8	200	21	6.4
10	250	25	7.5
12	300	28	8.5
14	350	30	9.1
16	400	35	11
18	450	39	12
20	500	42	13
24	600	46	14
30	750	57	17
≥36	**900**	**65**	**20**

* Support spacing recommendations are primarily governed by pipe diameter. However, pipe DR and joint design may decrease or increase recommended support spacing. For specific guidance, contact the pipe supplier.

† The recommended spacing should be reduced for sustained temperatures above 80°F (27°C). For specific guidance, contact the pipe supplier.

Source: Underground Solutions, Inc.

Figure 10-7 Example of short pipe sections being used to reduce drag friction

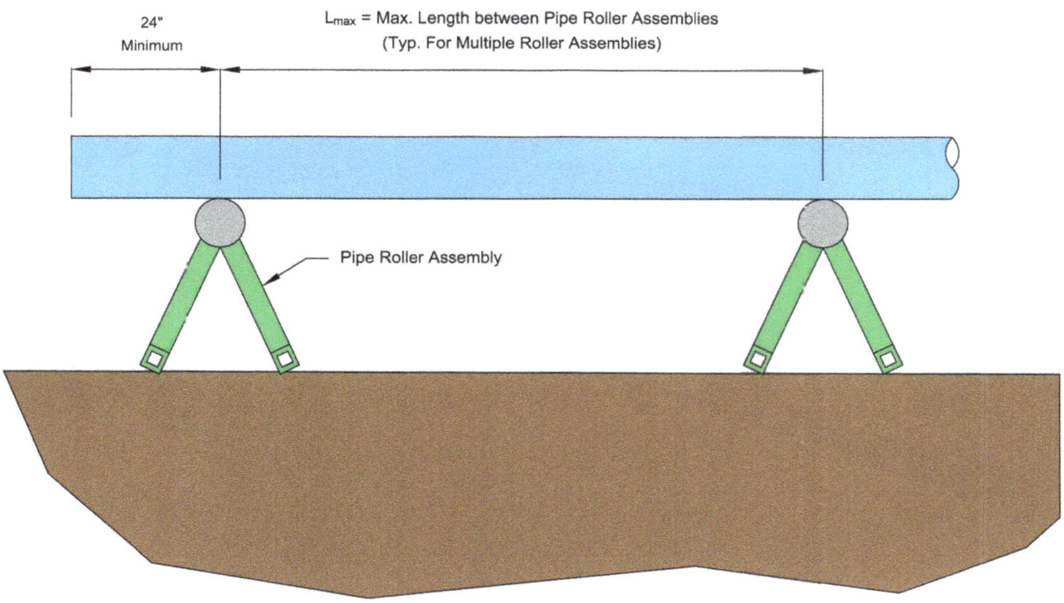

Source: Underground Solutions, Inc.

Figure 10-8 Typical roller placement

recommendations. A variety of means of roller insertion are depicted in Figures 10-9 through 10-12.

Pilot holes. The pilot hole is drilled with defined entry angle and exit angles. These angles are dependent on the minimum radius of curvature allowed within the right-of-way provided. The radius of curvature, at all locations during and after installation, must meet

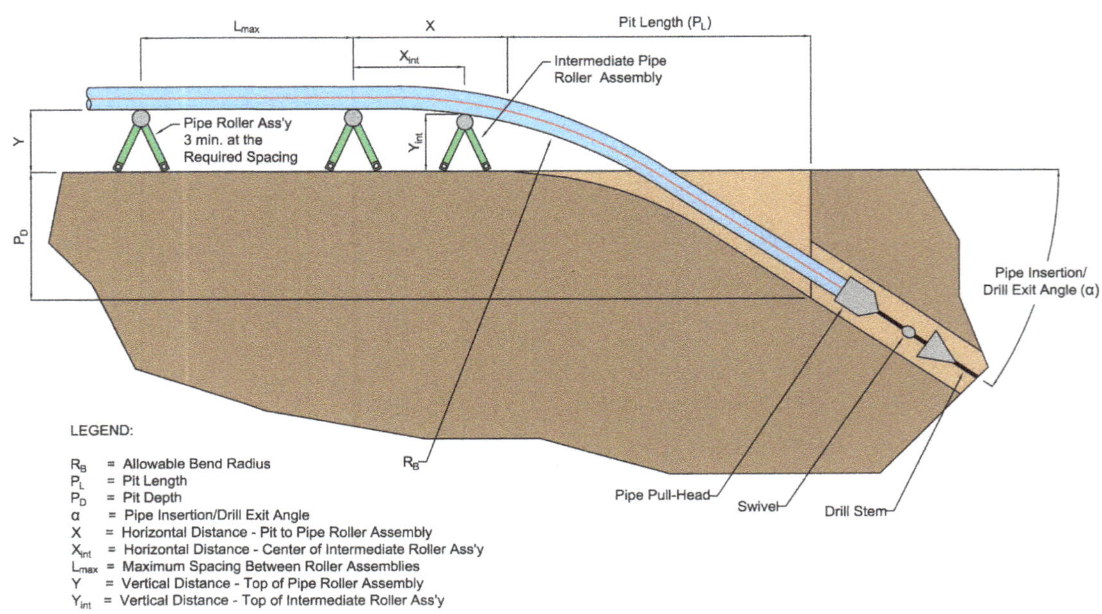

Source: Underground Solutions, Inc.

Figure 10-9 Insertion schematic for rollers

Source: Underground Solutions, Inc.

Figure 10-10 Field application of roller-supported insertion

or exceed the minimum allowable radius of curvature for the restrained-joint PVC pipe system used.

Reaming the pilot hole. Special cutters called reamers are successively pulled through the pilot hole to produce a bore large enough for installation of a pipe of the required diameter. Simultaneously, drilling fluid is pumped into the hole. The reamed hole allows drilling fluid to fill the annular space and to flow around the pipe.

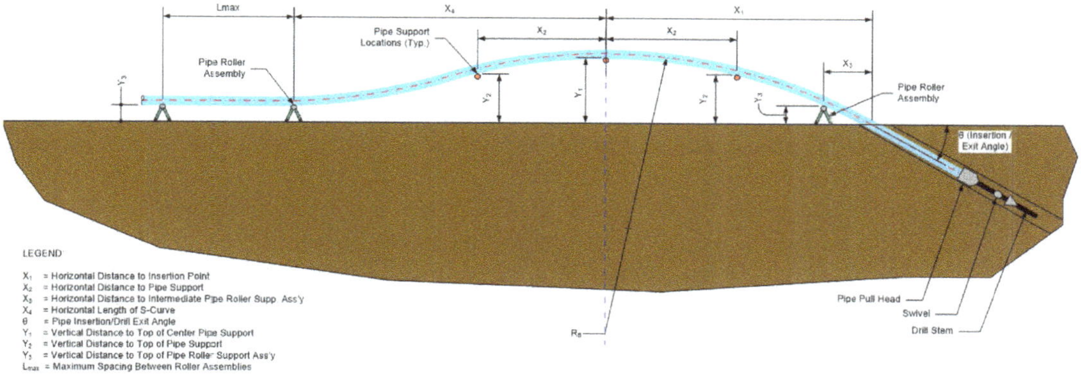

Source: Underground Solutions, Inc.

Figure 10-11 Schematic for an aerial insertion with rollers

Source: Underground Solutions, Inc.

Figure 10-12 Proper use of aerial slings, with rollers to reduce friction, for an HDD insertion

Borehole diameter. Soil type and water table may affect the borehole size. Generally, the diameter of the finished borehole should be as follows.

- For pipe sizes up to 24 in., the borehole should be 50 percent larger than the outside dimension of the pipe, pipe joint, pipe coupling, or external restraint—whichever is greatest.
- For pipe sizes larger than 24 in., the borehole should be no less than 12 in. larger than the outside dimension of the pipe, pipe joint, pipe coupling, or external restraint—whichever is greatest.

Drilling fluids. Drilling fluids lubricate the drill rod, drill head, reamer, and pipe. Ends of PVC pipe should be covered, capped, or plugged to prevent drilling fluid from entering the pipeline during installation. The drilling plan should include provisions to prevent the drilling fluid from escaping to the environment. When drilling under waterways with limited clearance, the frac-out potential of the bore and inadvertent surface return potential should be considered. The frac-out potential is not a pipe-driven parameter but is derived from soil type, depth, and drilling fluid pressure which can lead to inadvertent surface returns or future settlement.

Pullback. A pulling head, in combination with a swivel eye to prevent torsional stresses, connects the drilling rod or reamer to the leading end of the PVC pipe string. Pulling heads must comply with recommendations of the pipe manufacturer or the technology provider. The pullback operation should happen as soon as a reamed hole is completed to minimize the possibility of borehole collapse. Pullback forces can be monitored to ensure they remain within allowable limits. Safe pull (and push) forces vary with the type of joint used in combination with the pipe diameter and DR.

Pullback with butt-fused PVC pipe is usually performed in a continuous length. If necessary, intermediate fusions are performed during the pullback process. Segmented PVC pipe (pins and groove, splines and grooves, and grip ring joints) are usually strung out in long lengths or can be installed one joint at a time.

Trenchless Construction—Installation in Casings

A casing is the larger diameter pipe through which the smaller diameter carrier pipe is installed. When PVC pipe is installed under highways, runways, or railways, a casing may be required for the following reasons:

1. to prevent damage to structures due to soil erosion or settlement in the pipe zone,
2. to facilitate pipe removal and replacement in the future, or
3. to accommodate regulations or requirements imposed by public or private owners of property where pipe is installed.

PVC pipe may be used as the casing pipe and/or the carrier pipe. When PVC pipe is installed inside the casing pipe, casing spacers may be required to prevent damage to the pipe or pipe joints during installation and to provide proper long-term line support for belled end or mechanical jointed pipe. Maximum support spacing values for PVC pipe casing spacers should be limited to those given in Table 8.10 of the *Handbook of PVC Pipe Design and Construction*, 5th edition.

For pipe that will be pushed into position, casing spacers should be securely installed adjacent to the spigot-end insertion marks. Restrained joints suitable for compressive loads may be required if installation forces exceed the slippage resistance of casing spacers. For pipe that will be pulled into position, restrained joints having adequate allowable pull force resistance need to be used.

Casings are normally sized to provide an inside diameter at least 2 in. (50 mm) greater than the maximum outside diameter of the pipe joint, i.e., bell, casing spacer, or joint restraint device. Casing spacers also keep the pipe in position inside the casing. Various configurations are available to allow for a centered spacing or on a set distance from the invert of the casing. Care should be taken to provide adequate soil support for the carrier pipe just outside of the casing ends to avoid point loads from developing at the casing spacer contact points. For pipe joined by butt fusion with no external profile increase, laying the pipe on the invert of the casing where it is continuously supported is the preferred method.

Installing pipe through casings. After installation of the casing pipe, which is often bored or jacked into place itself, the restrained-joint PVC pipe can be pulled or pushed ("jacked") through casings. Installation should be done at a steady, controlled pace.

Annular space. The ODJ and OD of the carrier pipe is less than the inside diameter of the casing, which leaves an annular space between the two. In many cases, this annulus is filled with grout after installation. The method and material used for backfilling the annular space will affect the minimum clearance needed. In cases where the carrier pipeline may need to be removed from the casing, casing spacers can be used with the ends of the casing pipe sealed to prevent groundwater and soil from entering the annular space. See "Filling the Annular Space" at the end of this chapter.

Trenchless Rehabilitation—Sliplining

Sliplining is accomplished by inserting a lining pipe into a host pipe that needs rehabilitation. Unlike some other trenchless methods, the PVC lining pipe is fully structural, i.e., capable of carrying both the internal and external loads, and is not dependent upon the host pipe. PVC reliner pipe is usually pushed or pulled into the existing host pipe. Before installation, the host pipe is cleaned of debris and any service buildup (sediments and/or tuberculation), then surveyed for internal clearance, alignment, and obstructions.

In some cases, spot repairs are required on the host pipe in areas where the replacement pipe cannot pass. The largest PVC slipliner pipe that will fit is selected and inserted. Sizing of the slipliner pipe must take into account the maximum outside dimension of the reliner joints. For long insertion lengths, approximately 2 in. of clearance between the host pipe inside diameter and the maximum outside diameter of the reliner joints is recommended.

Hydraulic capacity can often be maintained despite the diameter reduction due to PVC pipe's low hydraulic friction, i.e., Hazen–Williams flow coefficient (C = 150) and Manning's 'n' = 0.009.

Loading. If portions of the PVC slipliner will extend beyond the host pipe the differences in loading conditions installation must be considered and accommodated in the design of the system. Loading conditions are not identical since there are differences in the two regions' ring deformation and longitudinal pipe deflection.

Annular space. Sliplining pipes are usually grouted in place. This prevents surface settlement that could be caused by the further deterioration of the host pipe. Installation material that approximates soil strength around the host pipe is recommended. See "Filling the Annular Space" later in this chapter.

Installation. There are two general methods used for sliplining installations:

1. *Segmental sliplining*—Liner is installed one pipe length at a time.
2. *Continuous sliplining*—Carrier pipe lengths are assembled, then installed in one long string or several strings with intermediate connections.

Segmental sliplining. With this method, PVC pipe may be assembled in segments at entry points along the length of the deteriorated host pipe. Pipe is inserted directly into the host pipe by either pulling or pushing. Joints capable of withstanding the pulling and/or pushing forces are required. Joint assembly shall be in accordance with manufacturer's instructions. Open-cut trenches are required to access the host pipe at strategic installation points.

Services need to be excavated, disconnected, and reconnected to the new PVC sliplining. Making connections to different materials also involves exposing both the lining pipe and the pipe to be connected with open cut excavation.

Continuous sliplining. In continuous sliplining, PVC carrier pipe is preassembled in long lengths before it is pulled into the deteriorated host pipe. The pull-in can be done in one unsegmented length or in sections that require an intermediate joint or butt-fusion. When required, taps are made after the new pipe is pulled into the host pipe.

The two primary components of the forces required for sliplining are:

1. Friction caused by the length of pipe being pulled above grade
2. Friction between the new pipe and the host pipe

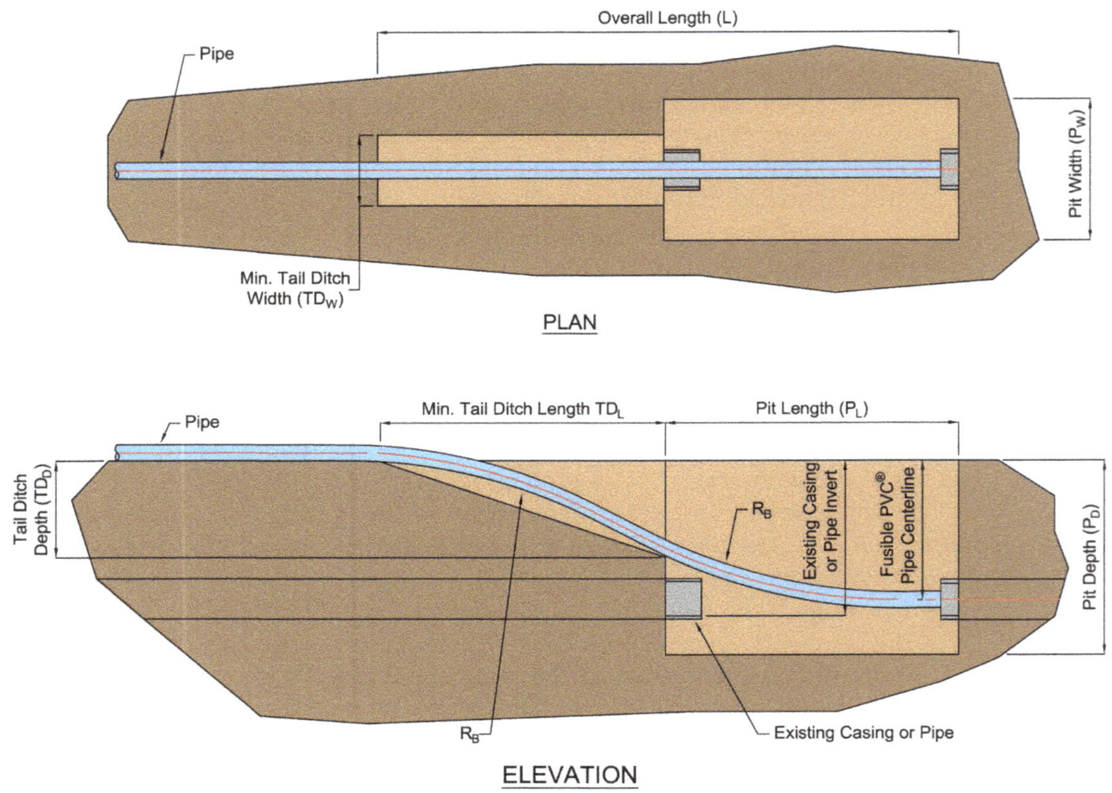

Source: Underground Solutions, Inc.

Figure 10-13 Schematic of slipline insertion pit pulled from grade

Installation pits must be excavated deep enough for PVC pipe installation to proceed, without the pipe being bent tighter than the minimum bending radius of the carrier pipe. In many cases, a significant amount of the alignment change for insertion can be accomplished above grade by supporting the pipe with rollers or equipment (Figure 10-13). This practice minimizes the length of pit needed. Pit length also dictates how much host pipe needs to be removed to allow for insertion.

Trenchless Rehabilitation—Pipe Bursting

In pipe bursting, an existing utility corridor is used to replace old pipe with completely new, fully structural pipe. Pipe bursting involves splitting or fracturing an existing pipeline and pushing that pipeline out into the surrounding soil, thus creating an area large enough for a new pipeline of the same or larger diameter to be pulled in.

Only joints designed for pulling are used for pipe bursting installations.

Size considerations. The diameter of the pipe being burst typically ranges from 2 in. to 30 in., although bursting has been used on pipes of larger diameter. Pipe bursting is commonly performed size-for-size or one size above the diameter of the existing pipe. Larger upsizing (upsizing by two or three sizes) has been done. With larger upsizing, there is greater potential for ground movement (upheaval) and adjacent utility disturbance. The larger the upsizing, the greater the force needed to burst the existing pipe and pull in the new pipe. Therefore, parameters and equipment need to be thoroughly defined when upsizing two to three sizes. The higher strength of PVC, relative to other thermoplastics, allows for a thinner wall, which minimizes upsizing requirements and/or provides for more flow area.

Pipe bursting methods. The recommended method of pipe bursting using PVC pipe is static bursting. In static bursting, a burst head that fractures or cuts the existing pipe is pulled through, followed by an expander head that pushes burst pipe and soil out of the alignment to allow the new pipe to follow behind the expander. The pull-in is continuous. PVC pipe is recommended for pipe bursting and widely used to replace concrete, cast iron, ductile iron, and asbestos cement pipe. Pneumatic bursting using a reciprocating hammer is not used with PVC pipe.

Trenchless Rehabilitation—Tight Fit Liners

Tight fit lining can be accomplished by expanding a specially formulated PVC pipe that has been inserted into a casing. The PVC liner pipe is brought to a tight fit inside the existing pipe through a combination of heat and pressure. These liners are generally referred to as "fold and form." Note that numerous fold and form PVC products exist and are specifically defined within ASTM product specification and installation guides and practices. Additionally, the manufacturers should be consulted for detailed design and installation information.

Before installation, the host pipe is cleaned of debris and service buildup. The line is then surveyed for alignment and obstructions. In some cases, spot repairs are required on the host pipe in areas where the replacement pipe would not be able to pass.

If the host pipe is a segmented pipe, the deflection of the joints and the lay length of the host pipe become design considerations. Joint spacing and deflection are used to determine the bend radius of the liner pipe that will be inserted.

PVC pipe's high Hazen–Williams flow coefficient (C = 150) and Manning's 'n' = 0.009, can often maintain or improve hydraulic capacity despite the slight reduction in diameter.

Filling the Annular Space

For sliplining or casing installations, the ODJ and OD of the new pipe is less than the ID of the host pipe or casing, which leaves an annular space between the two. In many cases, this annulus is filled with grout after installation, particularly if the host pipe is highly deteriorated (Figure 10-14). Grouting of the annular space provides additional support for the carrier pipe, helps protect the carrier pipe if the host pipe is in structural distress, and stops water infiltration through the host pipe or casing annular space.

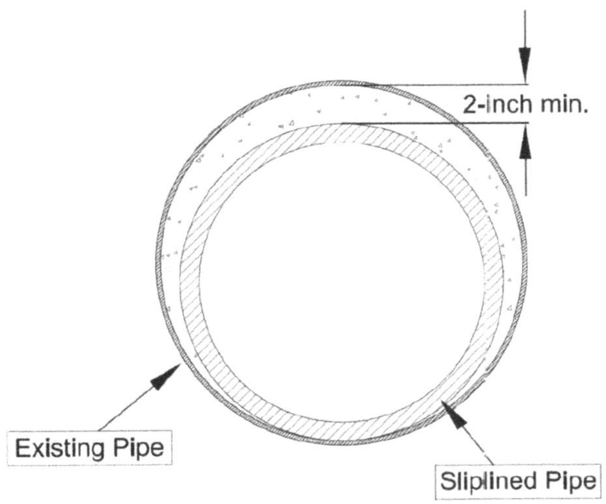

Source: Underground Solutions, Inc.

Figure 10-14 Annular space in sliplining

During the filling or grouting of the annular space, care must be exercised to keep the PVC pipe from floating out of its proper position. Spacers can be used for maintaining the PVC carrier pipe's position inside a casing or host pipe. If the carrier pipe is laid in the invert of the casing or host pipe, then filling with water can hold the pipe in its final alignment position. When the annular space between the casing and PVC carrier pipe is filled by pressure grouting, caution should be exercised to ensure that excess pressure does not cause the pipe to distort or collapse. Table 10-4 lists *critical grouting pressure* ($P_{critical}$) and *allowable grouting pressure* ($P_{allowable}$) as a function of the pipe dimension ratio (DR).

Table 10-4 Recommended grouting pressure limits

Dimension Ratio	$P_{critical}$, *psi (kPa)*	$P_{allowable}$, *psi (kPa)*
51	7.5 (51.7)	3.7 (26)
41	14.6 (101)	7.3 (50)
35	23.8 (164)	12 (83)
32.5	29.9 (206)	15 (103)
28	47.5 (328)	24 (165)
26	59.8 (412)	30 (207)
25	67.6 (488)	34 (235)
23.5	82.1 (566)	41 (283)
21	117 (807)	58 (400)
18	190 (1,310)	95 (655)
17	228 (1,572)	114 (786)
14	426 (2,937)	213 (1,470)
13.5	479 (3,300)	240 (1,650)

Notes:
1. $P_{allowable}$ is based on a safety factor against buckling of 2.0 (equal to DF = 0.5) at a temperature of 73°F, using E = 400,000 psi.
2. Grouting pressures must be reduced for increased wall temperatures per Table 10-4.

Table 10-5 Temperature corrections for modulus of elasticity, tensile load, and pressure

Temperature °F	Temperature °C	Adjustment Factor for Pressure, %	Adjustment Factor for Tensile, %	Adjustment Factor for Modulus, %
73.4	23	100	100	100
80	27	88	95	98
90	32	75	87	94
100	38	62	78	88
110	43	51	71	84
120	49	40	63	78
130	54	31	55	74
140	60	22	48	70

Notes:
1. Interpolate between the temperatures listed to calculate other correction factors.
2. The maximum recommended wall temperature for PVC pipe and fitting is 140°F (60°C) and for PVCO pipe 130°F (55°C).

In addition to controlling grout pressures, it is important to control grout placement and temperature. The values in Table 10-4 assume uniform grout pressures in the annulus. Distortion may result if the level of the grout becomes too great on one side of the pipe compared to the other. Using a higher stiffness (lower DR) pipe can be advantageous in this case as it offers greater resistance against distortion or collapse. If large amounts of grout—or grouts with a high heat of hydration temperature—are used, the heat of hydration must be taken into consideration as excess heat will lower pipe stiffness. Filling the carrier pipe with water will provide a heat sink and cooling to mitigate the effects of the heat of hydration. Temperature corrections for modulus of elasticity, tensile load, and pressure are provided in Table 10-5.

$$P_{\text{critical}} = \frac{2E}{\left(1 - v^2\right)\left(\text{DR} - 1\right)^3} \qquad \text{(Eq 10-1)}$$

where:

P_{critical} = critical buckling pressure, psi
E = modulus of elasticity of pipe material, psi
v = Poisson's ratio = 0.38 for PVC, dimensionless,
DR = dimension ratio = D_o/t, dimensionless,
D_o = pipe outside diameter, in.
t = pipe wall thickness, in.

Blown sand has also been used to fill the annular space and may facilitate removal of the carrier pipe in the future.

REFERENCE

Uni-Bell PVC Pipe Association. 2012 (5th ed.). *Uni-Bell Handbook of PVC Pipe—Design and Construction.* Dallas, Tex.: Uni-Bell PVC Pipe Association.

This page intentionally blank.

Chapter 11

Design for Thrust Restraint

When a pipeline is subjected to internal pressures, the resultant thrust forces on the pipe joints may cause joint separation. These thrust forces are a result of the same principles that cause a hydraulic cylinder to operate. Pressure acting on an area results in a force equal to the pressure of the fluid multiplied by the cross-sectional area over which the pressure acts ($F = PA$). In a hydraulic cylinder, the area is the cross-sectional area of the piston based on its outside diameter. In the case of push-on, bell-and-spigot joints in a PVC pipeline where the gasket is carried by the bell, the thrust force is the product of the line pressure and the cross-sectional area based on the pipe outside diameter at the joint.

In straight lengths of a pressurized, gasketed PVC pipeline, the thrust force at each joint is balanced by the thrust forces at adjacent joints. The weight of the pipe, backfill, and frictional resistance of the surrounding soil will generally be sufficient to prevent movement in straight lengths of pipe. Unbalanced forces, however, are developed at elbows, tees, wyes, reducers, valves, and dead ends. While unbalanced thrust forces are a result of internal pressure, one or more of the following conditions can affect the ultimate thrust:

- Transient pressure
- Fluid velocity
- Dead weight of pipe and liquid

Typically, the design of a thrust restraint system is based on the test pressure of the system. This assumes that the test pressure will be the highest sustained pressure the system will experience. Hydrodynamic forces from the fluid velocity are usually ignored as the hydrostatic, test pressure forces in a pipeline are generally greater than the hydrodynamic forces. Establishment of the design parameters and the selection of the appropriate method(s) of providing thrust restraint must be based on the present and future conditions.

As is the case for designing buildings, the design of thrust restraint systems should be conducted by an engineer with experience in water main construction as well as geotechnical engineering. All the unbalanced forces must be identified, quantified, and restrained in a safe and permanent manner. This chapter addresses basic thrust design and accepted restraint practices.

FORCES AT HORIZONTAL BENDS

One of the more common waterworks fittings is a horizontal bend. The resultant thrust at a horizontal bend is also the most frequent subject of joint restraint examination and discussion. At a bend, the forces on adjacent joints do not balance each other but combine to create a resultant force that tends to push the bend away from the pipeline. In the case of a horizontal bend, as illustrated in Figure 11-1 with a bend angle Δ, the resultant thrust is given by the following equation.

$$T = 2PA\sin(\Delta/2) \quad \text{(Eq 11-1)}$$

where:

T = resultant thrust force, lb
P = internal pressure, psi
A = cross-sectional area of the pipe based on outside diameter, in.^{2}
Δ = angle of deflection, degrees

As the angle Δ increases, the resulting thrust also increases. The thrust at a 90° bend is 1.8 times the thrust at a 45° bend and 3.6 times the thrust at a 22½° bend.

FORCES AT OTHER FITTINGS

As mentioned previously, thrust forces are also generated at other fittings as shown in Figure 11-2.

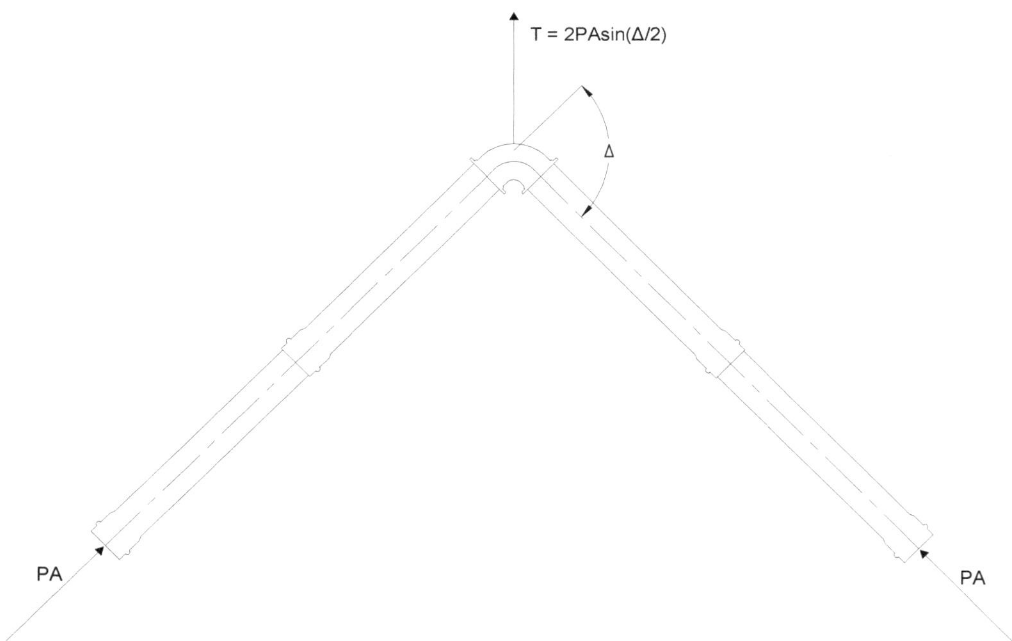

Figure 11-1 Free body diagram of forces at a pipeline bend

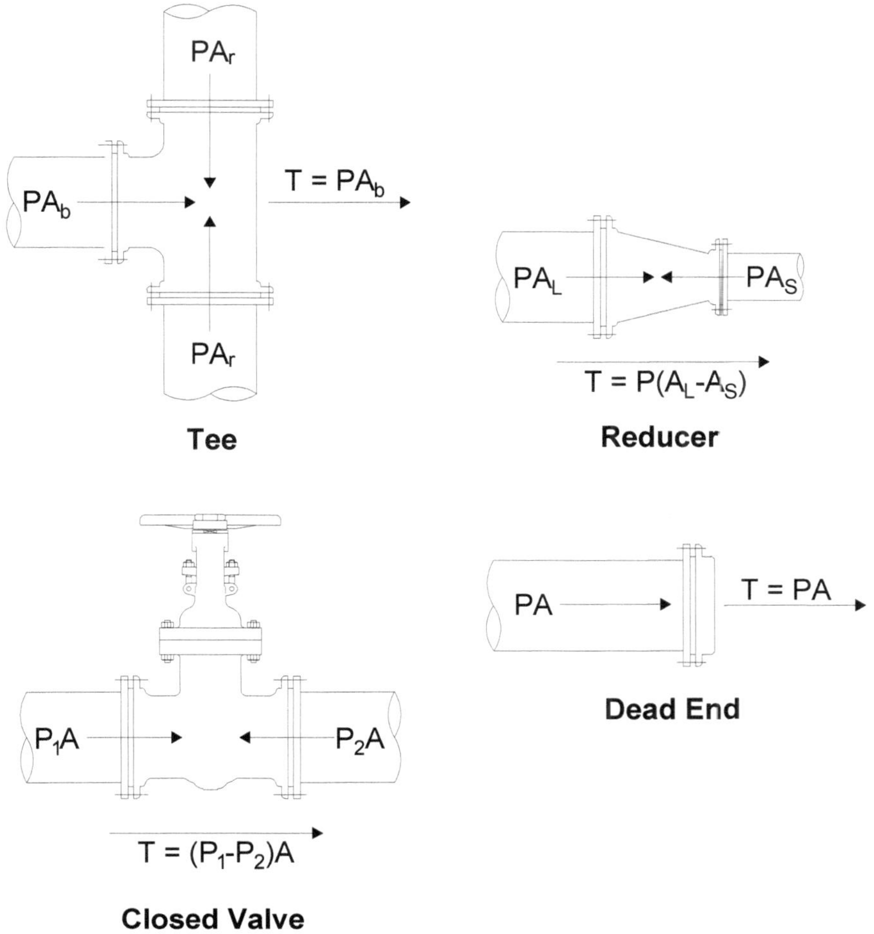

Figure 11-2 Thrust forces generated at other pipeline fittings

THRUST BLOCKS

One method of resisting the unbalanced forces in a pipeline is a thrust block. When properly designed and installed, a thrust block provides a satisfactory method of resisting the unbalanced forces in both a lateral and vertical direction.

Thrust blocks rely on the bearing strength of the soil behind the thrust block. With this method, the resultant thrust force in the pipeline is transferred through the thrust block to an area of compacted or undisturbed soil behind the thrust block that is large enough to safely resist the thrust loading. A typical thrust block for a horizontal bend is shown in Figure 11-3.

The size of the thrust block is related to the strength of the native soil: as the bearing strength of the undisturbed soil increases, the required size of the bearing area of the thrust block decreases (see Table 11-1).

The reliability of the thrust block may be compromised if digging occurs in the vicinity of the back of the thrust block in the future. Thrust blocks should be shown on the water main system plans, and a cautionary note about digging in the area should be on the drawing. In some cases, the required size of the block will render it impractical for use. This is particularly true in congested, urban areas. In poor soil conditions, large and heavy thrust blocks may cause settlement of the pipe. It can be difficult to avoid complete

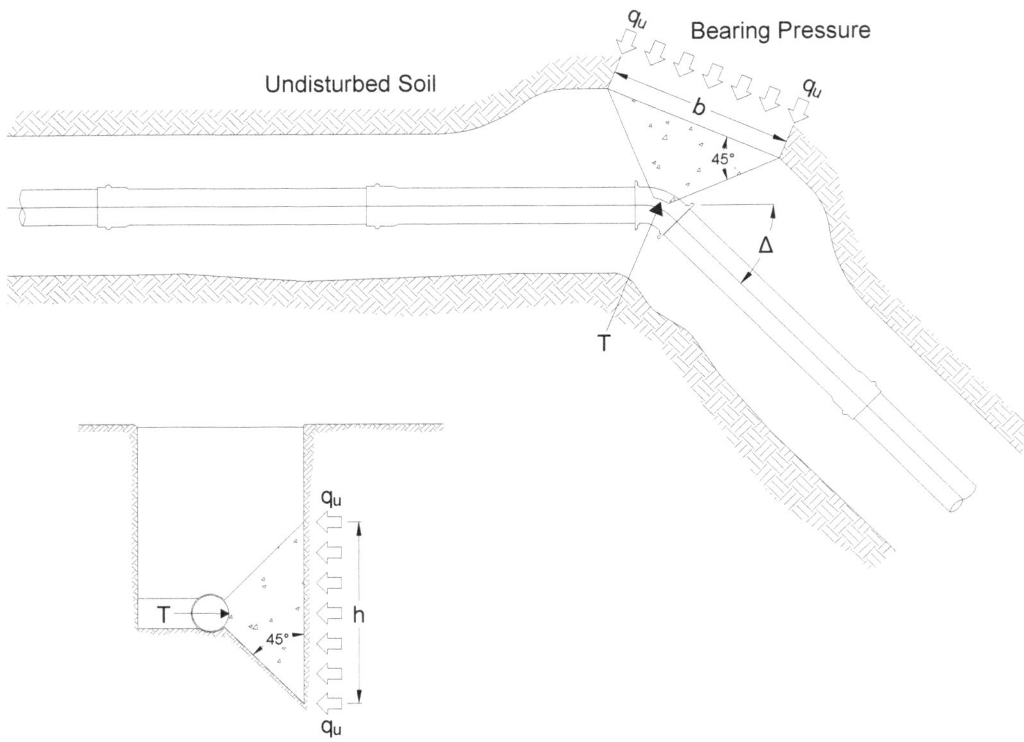

Figure 11-3 Typical thrust block for horizontal bend

Table 11-1 Estimated bearing strength of undisturbed soil

Soil Type	q_u, lb/ft^2
Muck, peat, etc.	0
Soft clay	500
Sand	1,000
Sand and gravel	1,500
Sand and gravel with clay	2,000
Sand and gravel, cemented with clay	4,000
Hard pan	5,000

Source: Adapted from Buried Pipe Design, Moser and Folkman, 2008.

encasement of compact fittings when poured-in-place thrust blocks are used. Additionally, with larger diameter pipelines the extreme volume of the required thrust blocks increases the complications already mentioned.

Thrust Block Design

The design and installation of thrust blocks is a critical part of a successful water main project. It is important that the soil-bearing capacity be carefully and properly assessed. The design engineer is responsible for determining the safe bearing strength of the undisturbed native soil. It is recommended that a qualified geotechnical engineer be engaged in that process. The present and possible future water table elevation will influence the soil-bearing capacity and must also be considered.

Additional design criteria for thrust blocks are as follows.

- The thrust block bearing surface should be placed against undisturbed soil or compacted backfill.
- The vertical height of the thrust block should be less than half the total depth to the final ground surface.
- The width of the thrust block should be at least two times the height of the thrust block.
- The maximum operating pressure, transient pressure, and test pressure must be considered in computing the unbalanced force.

The required thrust block bearing area can be calculated using the following equation.

$$A = (T/q_u) \times SF \tag{Eq 11-2}$$

where:

A = bearing area of thrust block, ft^2
T = resultant thrust force, lb
q_u = soil bearing capacity, lb/ft^2
SF = safety factor (suggested minimum, 1.5)

*This design process is based on estimated bearing strength of undisturbed soil and a pipe burial depth of at least 2 ft.

Some typical soil bearing capacity values are presented in Table 11-1.

The resultant thrust force for the horizontal bend shown in Figure 11-3 is determined using Eq 11-1. Figure 11-4 shows examples of other fitting and thrust block placements.

As indicated in Figure 11-2, the resultant thrust forces used to calculate the size of the thrust block engagement area for the fittings shown in Figure 11-4 are as follows.

Dead-end thrust forces develop at dead ends, the branch of a tee, and a closed valve. That thrust force can be determined with Eq 11-3.

$$T = PA \tag{Eq 11-3}$$

where:

T = resultant thrust force, lb
P = internal pressure, psi
A = internal area based on the diameter of the sealing element, $in.^2$

In a similar fashion, the resultant thrust force from the large side of a reducer is partially offset by the resultant thrust force of the small size. Equation 11-4 can be used to determine that thrust force.

$$T = P(A_L - A_S) \tag{Eq 11-4}$$

where:

T = resultant thrust force, lb
P = internal pressure, psi
A_L = sealing element area on the large side of the reducer, $in.^2$
A_S = sealing element area on the small side of the reducer, $in.^2$

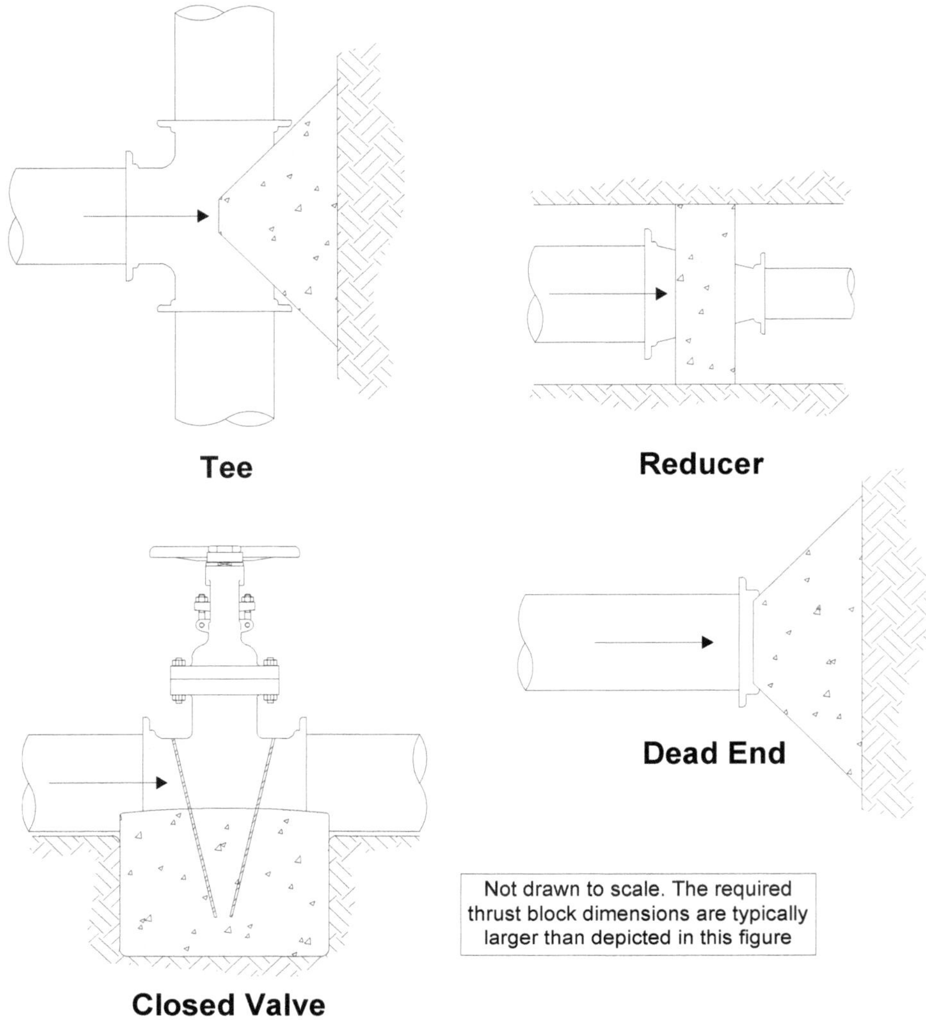

Figure 11-4 Thrust block placement to resist thrust forces

Thrust forces that are not directed into the surrounding soil (such as the upward thrust force generated at a vertical down bend) must be countered by the weight of the gravity thrust block as shown in Figure 11-5.

The vertical component of the thrust force for this fitting can be calculated using Eq 11-5.

$$V_g = (SF \times PA\sin\Delta) / W_{tb} \quad \text{(Eq 11-5)}$$

where:

V_g = volume of the gravity thrust block, ft^3
SF = safety factor (suggested minimum, 1.5)
P = internal pressure, psi
A = cross-sectional area of the pipe outside diameter, $in.^2$
Δ = angle of deflection, degrees
W_{tb} = unit weight of the thrust block material, lb/ft^3

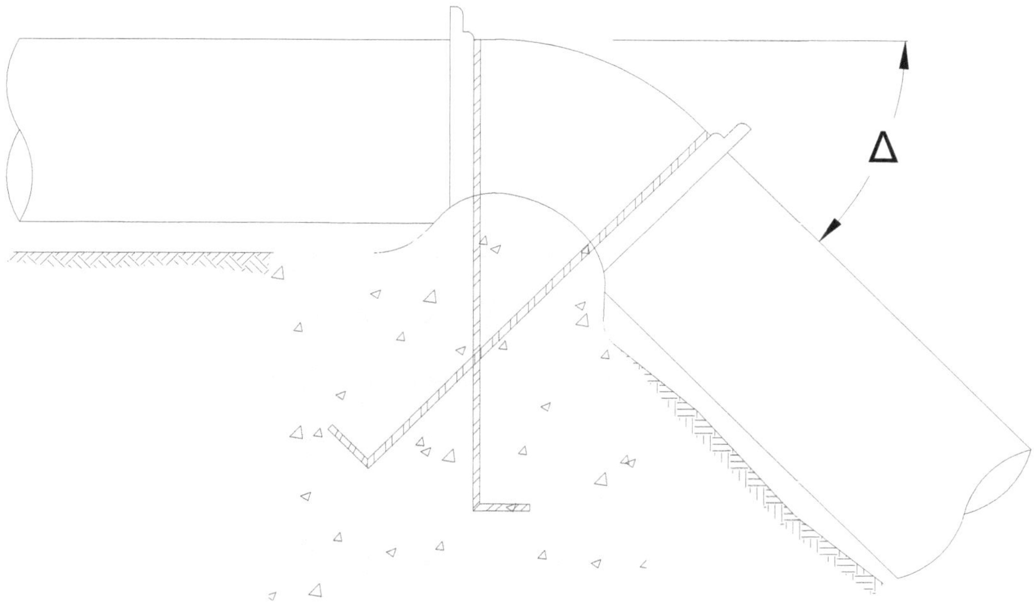

Figure 11-5 Gravity thrust block for a vertical down bend

The horizontal component is resisted by the bearing strength of the soil in the horizontal thrust direction with a thrust force determined by Eq 11-6.

$$T_x = PA(1 - \cos\Delta) \quad \text{(Eq 11-6)}$$

where:

T_x = horizontal thrust force
P = internal pressure, psi
A = cross-sectional area of the pipe based on outside diameter, in.^{2}
Δ = angle of deflection, degrees

RESTRAINED JOINTS

By restraining certain joints along the pipeline, the resultant thrust force is transferred to the surrounding soil by the pipeline itself. In a properly designed pipeline using restrained joints, the bearing strength and/or the frictional resistance of the soil balance the thrust forces.

There is no requirement to build forms, keep the trench open, or delay proof testing of the pipeline while the thrust blocks cure. In crowded urban areas, the use of restrained joints can enable the trench to be opened, the pipe and fitting installed, and the trench closed, all in the same day. Experience has shown that in a piping system with joints that that are properly restrained, concrete thrust blocks can be entirely eliminated. However, some jurisdictions may require the installation of thrust blocks in addition to restrained joints.

Only joint restraint devices manufactured and tested for use with PVC pressure pipe should be considered.*

* Some restraint products are manufactured specifically for C909 PVC or C900 PVC and are not interchangeable. It is important to verify manufacturer's recommendations.

Major Division			Group Symbols	Typical Names	Classification Criteria		
COARSE-GRAINED SOILS More than 50% retained in No. 200 sieve*	GRAVELS 50% or more of coarse fraction retained on No. 4 sieve	CLEAN GRAVELS	GW	Well-graded gravels and gravel-sand mixtures, little or no fines	Classification on basis of percentage of fines Less than 5% Pass No. 200 sieve: GW, GP, SW, SP More than 12% Pass No. 200 sieve: GM, GC, SM, SC 5% to 12% Past No. 200 sieve: Borderline Classification requiring use of dual symbols	$C_x = D_{60} / D_{30}$ Greater than 4 $C_z = (D_{30})^2 / D_{10} \times D_{60}$ Between 1 and 3	
			GP	Poorly graded gravels and gravel-sand mixtures, little or no fines		Not meeting both criteria for GW	
		GRAVELS WITH FINES	GM	Silty gravels, gravel-sand-silt mixtures		Atterberg limits plot below "A" line or plasticity index less than 4.	Atterberg limits plotting in hatched areas are borderline classifications requiring use of dual symbols.
			GC	Clayey gravels, gravel-sand-clay mixtures		Atterberg limits plot below "A" line or plasticity index less than 7.	
	SANDS More than 50% of coarse fraction passes No. 4 sieve	CLEAN SANDS	SW	Well-graded sands and gravelly sands, little or no fines		$C_x = D_{60} / D_{30}$ Greater than 6 $C_z = (D_{30})^2 / D_{10} \times D_{60}$ Between 1 and 3	
			SP	Poorly graded sands and gravelly sands, little or no fines		Not meeting both criteria for SW	
		SANDS WITH FINES	SM	Silty sands, sand-silt mixtures		Atterberg limits plot below "A" line or plasticity index less than 4.	Atterberg limits plotting in hatched areas are borderline classifications requiring use of dual symbols.
			SC	Clayey sands, sand-clay mixtures		Atterberg limits plot below "A" line or plasticity index less than 7.	
FINE-GRAINED SOILS 50% or more passes No. 200 sieve*	SILTS AND CLAYS Liquid limit 50% or less		ML	Inorganic silts, very fine sands, rock flour, silty or clayey fine sands	PLASTICITY CHART For classification of fine-grained soils and fine fraction or coarse-grained soils. Atterberg limits plotting in hatched area are borderline classifications requiring use of dual symbols. Equation of A-line:		
			CL	Inorganic clays of low to medium plasticity, gravelly clays, sandy clays, silty clays, lean clays			
			OL	Organic silts and organic silty clays of low plasticity			
	SILTS AND CLAYS Liquid limit greater than 50%		MH	Inorganic silts, micaceous or diamaceous fine sands or silts, clastic silts			
			CH	Inorganic clays of high plasticity, fat clays			
			OH	Organic clays of medium to high plasticity			
Highly Organic Soils			PT	Peat, muck, and other highly organic soils			

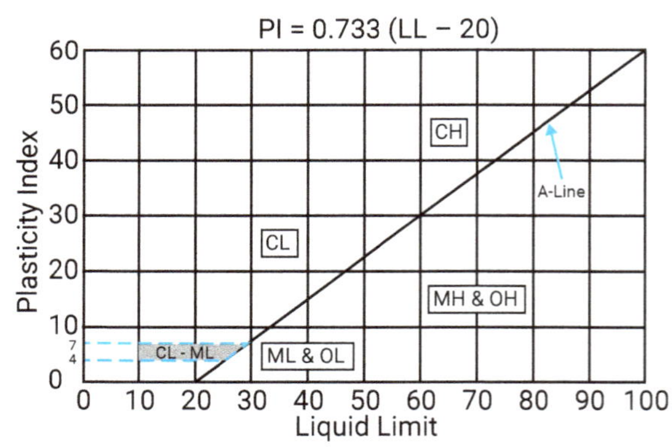

* Based on a material passing the 3-in. (75-mm) sieve

Source: ASTM International

Figure 11-6 The Unified Soil Classification System

Table 11-2 Pipe to soil interface properties used to calculate R_s and F_s

						K_n Trench Type		
Soil Group	f_c	$C\ (lb/ft^2)$	$f\phi$	$\phi\ (deg)$	$\gamma\ (lb/ft^3)$	3	4	5
GW & SW	0	0	0.7	35	110	0.60	0.85	1.00
GP & SP	0	0	0.7	31	110	0.60	0.85	1.00
GM & SM	0	0	0.6	30	110	0.60	0.85	1.00
GC & SC	0.2	225	0.6	25	100	0.60	0.85	1.00
GL	0.3	250	0.5	20	100	0.60	0.85	1.00
ML	0	0	0.5	29	100	0.60	0.85	1.00

Balancing the Resultant Thrust Force

The basic forces for restrained joint design are the bearing resistance "R_s" and the frictional resistance "F_s". The values of R_s and F_s are dependent upon

- classification of the native soil and bedding material,
- pipe burial depth, and
- compaction of the pipeline embedment and backfill.

Soil can be classified in accordance with The Unified Soil Classification System detailed in ASTM D2487; the soil classes are summarized in Figure 11-6.

Trench compaction details are shown in Figure 7-2. Because of the low compaction trench types 1 and 2 may allow for too much movement of the pipe into the soil and are not recommended for restrained joint systems.

Finally, the resistance that is generated along the required restrained length is reliant on friction at the interface between the soil and the pipe. Tests were made on actual PVC pipe surfaces embedded in a variety of soils to determine actual values for the pipe-to-soil interface.*

The properties shown in Table 11-2 are used in determining the values of R_s and F_s. The formulas for calculating R_s and F_s are provided in Eq 11-8 and Eq 11-11.

Note that soils in the CL and ML groups must be monitored closely because moisture content is difficult to control during compaction. Free-draining soils are much better pipe bedding material. Soils in the MH, CH, OL, OH, and PT groups are not recommended for restrained pipe bedding.

Pipelines laid in highly plastic soils subject to high moisture content are usually bedded in some type of granular material. In cases where the bedding material has a higher bearing value than the native soil, the value of F_s should be calculated using the bedding material values, while the value of R_s should be based on the native soil. In this case, the undrained shear strength values should be used for cohesion according to the "Phi = 0" principle. Various field tests (such as the vane shear test, ASTM D2573) or laboratory tests (such as the unconfined compression test, ASTM D2166) can be used to provide values of the undrained shear strength. Standard Penetration Test results (ASTM D1586) can also be used to approximate undrained shear strength results using published tables found in many geotechnical design manuals.

* Kennedy, Harold Jr.; Shumard, Dennis and Cary, M. 1989. Investigation of Pipe-to-Soil Friction and Its Effect on Thrust Restraint Design for PVC and Ductile Iron Pipe, EBAA Iron Inc., Presented at AWWA Distribution Systems Symposium, September 1989.

Table 11-3 In-situ values of fine-grained soils that are used to calculate R_s

Soil Group*	$C = S_u$ (lb/ft^2)	γ (lb/ft^3)	K_n Trench Type 3	K_n Trench Type 4	K_n Trench Type 5
CL	450	100	0.60	0.85	1.00
CH	400	100	0.40	0.60	0.85
ML	300	100	0.60	0.85	1.00
MH	250	100	0.40	0.60	0.85

Source: ASTM International
* Soil group per ASTM D2487.

The values in Table 11-3 are for near saturated, undisturbed soils, type CL, ML, CH, and MH, with the pipe surrounded with sands or gravels having a minimum compaction of 80 percent, determined using the standard Proctor test (ASTM D698). While these values are conservative for most situations, a competent soils engineer should be engaged for pipelines in wetlands and river bottoms.

Bearing Resistance, R_s

The passive resistance of the soil is generated at the pipe as the pipe tries to move into the surrounding soil. The maximum resistance to this movement can be calculated with the Rankine Passive Pressure formula. The amount of movement required to generate the resistance depends on the compressibility of the soil. In general, soils having a standard proctor density of 80 percent or greater require very little movement to generate the maximum passive resistance of the soil. Because the compressibility of the soil can vary greatly between the suggested trench types shown in Figure 7-2, the design value of passive pressure should be modified by an empirical constant, K_n, to ensure that excessive movement does not occur. Values of K_n vary by trench and material type and can be found in Tables 11-2 and 11-3. The number of restrained joints required can be minimized by specifying trench type 4 or 5 shown in Chapter 7, Figure 7-2.

Rankine's passive pressure formula is given by

$$P_p = \gamma H_c N_\varphi + 2CQ \quad \text{(Eq 11-7)}$$

where:

P_p = passive pressure of the soil, lb/ft^2
γ = unit weight of soil (unit weight of backfill for loose soil, native soil unit weight for compacted bedding), lb/ft^3
H_c = mean depth from surface to plane of resistance (centerline of pipe), ft
φ = internal friction angle of the soil
C = cohesion of the soil, lb/ft^2
N_φ = $\tan^2(45° + \varphi/2)$
Q = $\tan(45° + \varphi/2)$

Therefore, the bearing resistance on a pipe is defined as

$$R_s = K_n P_p D \quad \text{(Eq 11-8)}$$

where:

R_s = unit length bearing resistance, lb/ft
P_p = passive pressure of the soil, lb/ft^2
K_n = trench compaction factor (see Tables 11-2 and 11-3)
D = outside diameter of the pipe, ft

Frictional Resistance, F_s

The frictional force acting to oppose movement of the pipeline is a function of the internal shear strength of the soil in relation to the roughness of the pipe's surface.

Internal shear strength of a soil can be expressed by the Coulomb equation:

$$S = C + \sigma \tan(\varphi) \qquad \text{(Eq 11-9)}$$

where:

- S = shear strength of the soil, lb/ft^2
- C = cohesion of the soil (zero intercept of a plot of shear strength versus normal stress), lb/ft^2
- σ = normal stress (acting perpendicular to the plane of shearing), lb/ft^2
- $\tan \varphi$ = slope of the straight-line plot of shear strength versus normal force
- φ = angle of internal friction of the soil

Potyondy performed a series of investigations to study shearing resistance (skin friction) at the soil/surface interface for various construction materials in contact with different soils. His conclusions define the shearing resistance at the soil/surface interface in terms of a variation in the Coulomb equation using modified values of cohesion and the internal friction angle φ. Potyondy redefined these values as constants for a particular material/soil interface within a given moisture range.

Potyondy's equation for the shear strength of a soil/material surface interface is as follows:

$$S_p = f_c C + \sigma \tan(f_\varphi \varphi) \qquad \text{(Eq 11-10)}$$

where:

- S_p = shearing strength of the surface-to-soil interface, lb/ft^2
- f_c = proportionality constant relating the cohesion of a series of direct shear tests on the surface-to-soil interface and the cohesion intercept of the soil
- f_φ = proportionality constant relating friction angle of a direct shear series (surface/soil) interface to φ

Potyondy's study was based on surfaces commonly used for piles (e.g., concrete, wood, smooth and rusted steel). In the early uses of Potyondy's work for the design of pipeline restraint systems, the assumption was made that pipe surfaces were roughly equivalent from a frictional standpoint to rusted steel surfaces, and therefore constants of proportionality recommended by Potyondy for these surfaces were used. However, tests that were made on actual PVC pipe surfaces indicate that values chosen for the original design were not applicable to PVC pipe. It was also found in these studies that the values of f_c and f_φ vary with moisture content and the plasticity index of each soil. The selection of f_c and f_φ for use in this manual are based on the tests on actual pipe surfaces and are provided in Table 11-2 using trench conditions illustrated in Figure 11-7. Applying Potyondy's concepts, the pipe-to-soil friction is defined as follows:

$$F_s = A_p (f_c C) + W \tan(f_\varphi \varphi) \qquad \text{(Eq 11-11)}$$

where:

- F_s = unit length frictional resistance based on half the pipe circumference, lb/ft
- A_p = area of pipe surface bearing against the soil (one-half the pipe circumference, ft^2
- f_c = pipe to soil cohesion modifier
- C = cohesion of the soil
- W = normal force per unit length, lb/ft = $2W_e + W_p + W_w$

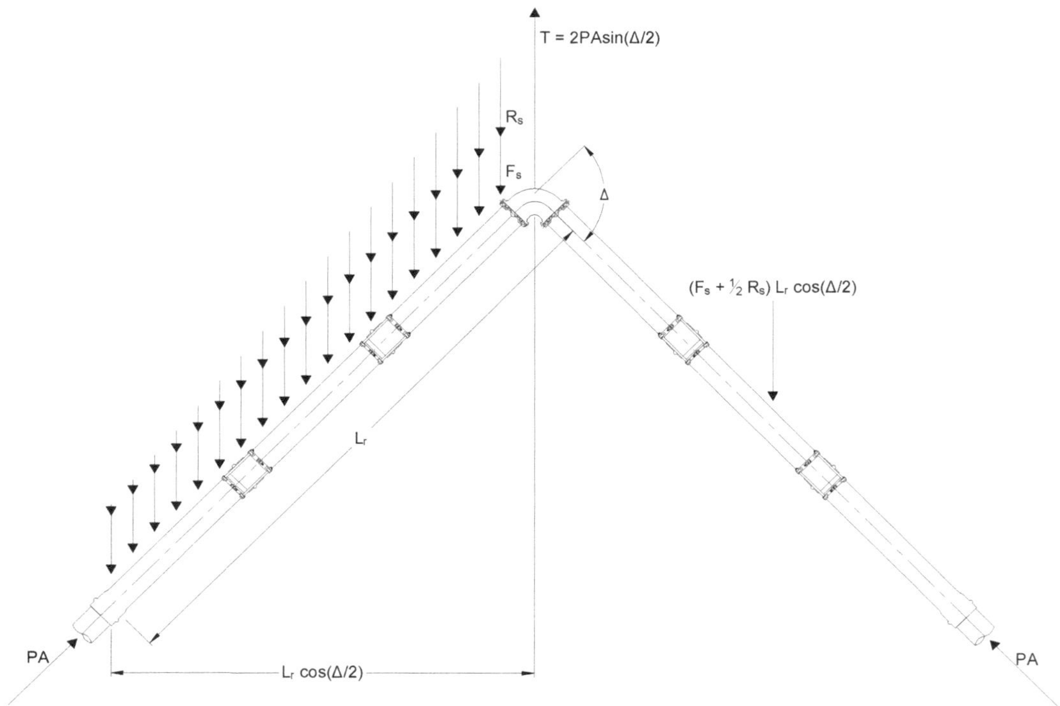

Figure 11-7 Bearing and frictional resistance acting on a restrained horizontal bend

$2W_e$ = vertical load on top and bottom surfaces of the pipe taken as the prism load, lb/ft

$W_p + W_w$ = weight of pipe plus weight of water, lb/ft

f_φ = pipe to soil friction angle modifier

φ = angle of internal friction of the soil

Equation 11-11 is for use with bends and uses friction acting on half of the pipe circumference. The following equation is to be used for dead ends and reducers where friction acts on the entire circumference of the pipe.

$$F_{sb} = A_{pb}(f_c C) + W\tan(f_\varphi \varphi) \qquad \text{(Eq 11-12)}$$

where:

F_{sb} = unit length frictional resistance based on the full pipe circumference, lb/ft

A_{pb} = area of pipe surface bearing against the soil (full pipe circumference, ft^2)

f_c = pipe to soil cohesion modifier

C = cohesion of the soil

W = normal force per unit length, lb/ft = $2W_e + W_p + W_w$

$2W_e$ = vertical load on top and bottom surfaces of the pipe taken as the prism load, lb/ft

$W_p + W_w$ = weight of pipe plus weight of water, lb/ft

f_φ = pipe to soil friction angle modifier

φ = angle of internal friction of the soil

For design purposes, friction values should always be based on the soil used for bedding. Native soil friction values should be used only when the native soil is also the bedding material.

Restrained Length Design for Horizontal Bend

Starting with a horizontal bend as an initial example, the bearing resistance, R_s, and frictional resistance, F_s, act on the pipe surfaces that are being "forced" into the soil as the pipe assembly attempts to move in the direction of the resultant thrust. The actual movement into the soil is only slight.

It is important to note that as the horizontal bend assembly is "pushed" into the soil only half of the pipe surface engages the resisting soil. The other half is "moving away" from the soil. Consequently, the resistive frictional forces are acting on only half the circumference of the pipe.

Figure 11-7 is a diagram of the resistive forces distributed along a restrained length "L" of the bend. The resistive forces occur on the restrained lengths of pipe on both sides of the horizontal bend. R_s is assumed to be a maximum at the bend and decrease to a value of zero at a distance of "L" away from the fitting along the leg on each side. F_s is assumed to act uniformly along L and act only on the outer half circumference of the pipe.

As shown above, the following equation can be used to determine the required restrained length, L.

$$L = SF \times PA\tan(\Delta/2)/\left(F_s + \frac{1}{2}R_s\right) \qquad \text{(Eq 11-13)}$$

where:

- SF = safety factor (suggested minimum, 1.5)
- P = internal pressure, psi
- A = cross-sectional area of the pipe based on outside diameter, in.^{2}
- Δ = angle of deflection, degrees
- F_s = pipe to soil friction based on half the circumference of the pipe, lb/ft
- R_s = bearing resistance of the soil along the pipe, lb/ft

For this fitting application, and those that follow, this design approach requires that all pipe joints within the distance "L" from the fitting be restrained. Any joint outside of the length "L" does not need to be restrained.

Restrained Length Design for Dead Ends

The forces acting on a dead-end length of pipe are shown in Figure 11-8. The thrust restraint design for dead ends is also useful in the determination of the restrained length requirement for several other fittings.

For a dead end, the frictional resistance is acting on the entire circumference as the pipe attempts to move into the soil in the direction of the resultant thrust force, T. This full-circumference frictional resistance is noted to be F_{sb}. The bearing resistance at a dead end is treated as insignificant so the equation for determining the restrained length is

$$L = SF \times PA/F_{sb} \qquad \text{(Eq 11-14)}$$

where:

- SF = safety factor (suggested minimum, 1.5)
- P = internal pressure, psi
- A = cross-sectional area of the pipe based on outside diameter, in.^{2}
- F_{sb} = unit frictional resistance based on the entire pipe circumference, lb/ft

Because there is no bearing resistance used in the calculations, "L" can be quite long for a dead end.

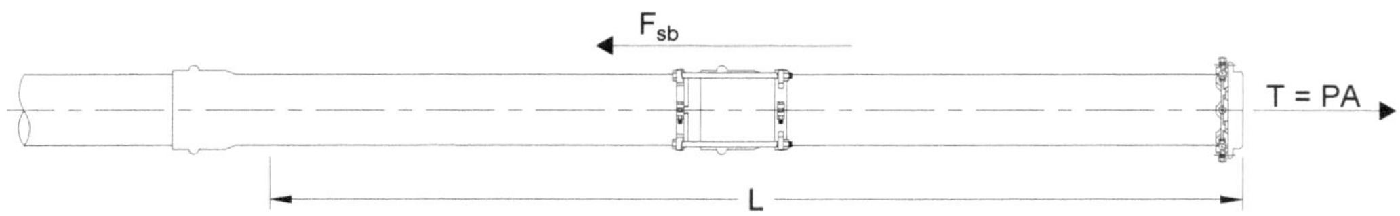

Figure 11-8 Restraint design for a dead end

Restrained Length Design for In-Line Valves

When closed, a valve in a pipe system acts as a termination point. Therefore, valves need to be treated as dead ends. The pressurized side of the valve must be restrained in the same manner as the dead end discussed previously, especially if the valve is closed during proof testing of the system. If the valve will only be closed during the operation of the pipeline, the restrained length may be based on the operating pressure of the system in order to reduce the restrained length requirement. Also, if the valve may be pressurized on either one side or the other, the pipe on each side of the valve will need to be restrained accordingly. This is especially important when a valve is closed so that work can be performed on the non-pressurized side of the valve. If the valve is not restrained properly and a pipe joint on the pressurized side of the valve separates, severe property damage, physical harm, or even death can occur.

Restrained Length Design for Reducers

The resultant thrust force for a reducer is similar to that of a dead end acting on the large side of the reducer with the exception that a countering thrust force is developed on the small side of the fitting. This is shown in Figure 11-9.

The countering thrust force from the small side reduces the restrained length on the large side when compared to a dead end. The recommended restraint length equation for a reducer is as follows:

$$L_L = SF \times P(A_L - A_S)/(F_{sb})_L \qquad \text{(Eq 11-15)}$$

where:

- SF = safety factor (suggested minimum, 1.5)
- L_L = restrained length on the large side of the reducer, ft
- A_L = cross sectional area of pipe on the large side of the reducer, in.^{2}
- A_S = cross sectional area of pipe on the small side of the reducer, in.^{2}
- $(F_{sb})_L$ = unit frictional resistance based on the entire pipe circumference, large side of the reducer, lb/ft

In a similar manner to that of the in-line valve, it is important to restrain the length, prescribed by Eq 11-15, on the large side of the reducer in order to prevent the pipe joints on the small side of the reducer from being over-inserted.

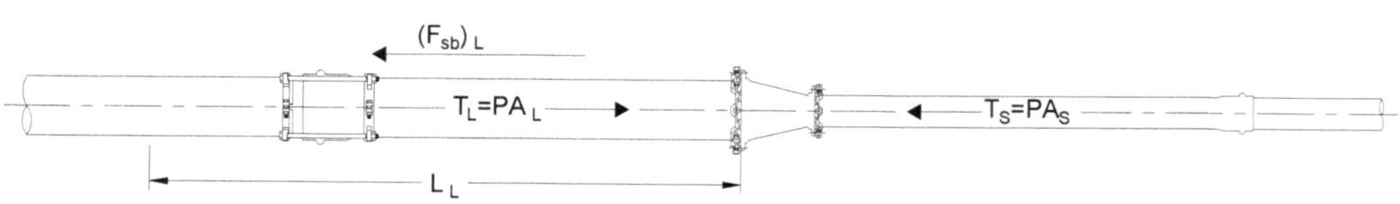

Figure 11-9 Restrained length design for a reducer

Restrained Length Design for Tees

With thrust blocks, the dead-end thrust force is transferred to the native soil behind the branch of the tee as shown in Figure 11-4. In a restrained system, a tee is a bit of a hybrid. The resultant thrust force that must be addressed occurs at the dead-end branch of the tee. As with the horizontal bend, bearing resistance develops on the back side of the tee much like a thrust block. During the design process, the designer must designate a minimum pipe length going into and out of the run of the tee. Doing so provides a fixed amount of bearing resistance at the back side of the tee based on that minimum length of pipe. The bearing is at a maximum at the tee and decreases to a value of zero at L_r away from the tee. This is shown in Figure 11-10.

This fixed amount of bearing resistance behind the tee is not reliant upon the pressure in the line nor the size of the branch of the tee. It is based solely on the diameter of the pipe along the run and the length of the pipe going into and out of the run of the tee. If the thrust force at the branch is smaller than the resisting force behind the tee, then all that is needed for each of the joints of the tee be restrained. If the force at the branch is larger than the resisting force behind the tee, then a length of pipe along the branch will need to be restrained in order to add enough frictional resistance along the branch to safely balance the thrust force at the branch. The equation used to determine the proper restrained length of piping at the branch is

$$L_b = SF \times (PA_b - R_S L_r)/F_{sb} \qquad \text{(Eq 11-16)}$$

where:

L_b = required restrained length along the branch of the tee, ft
SF = safety factor (suggested minimum, 1.5)
P = internal pressure, psi
A_b = cross sectional area of the branch pipe, in.^{2}
R_s = Unit bearing resistance of the run pipe, lb/ft
L_r = minimum pipe length going into and out of the run of the tee, ft
F_{sb} = unit frictional resistance based on the entire branch pipe circumference, lb/ft

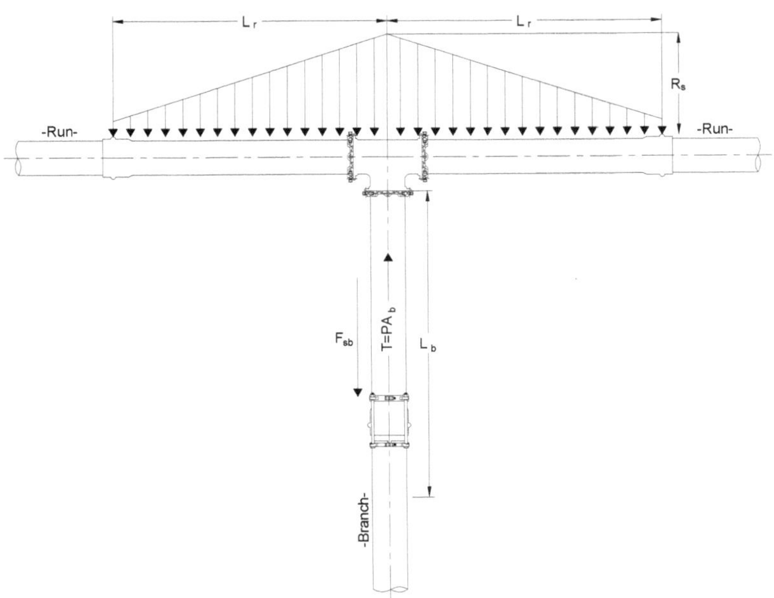

Figure 11-10 Restrained length design for a tee

Restrained Length Design for Hydrant Lateral

For a typical hydrant lateral, there is a tee, a valve, and a hydrant. Depending on the layout, there may be a pipe length between the branch of the tee and the valve and there may also be a pipe length between the valve and the shoe of the hydrant. In a restrained system, all joints in the hydrant lateral are restrained and thus no thrust force is developed.

Restrained Length Design for Vertical Down Bends

Because the soil above a vertical down bend is disturbed, it should not be relied upon for bearing resistance against upward thrust. Therefore, the calculation for the restrained length ignores bearing resistance. As a result, the restrained length for a vertical down bend, Figure 11-11, is significantly greater than that for a horizontal bend.

The equation for a vertical down bend is

$$L = SF \times PA\tan(\Delta / 2)/F_s \qquad \text{(Eq 11-17)}$$

where:

SF = safety factor (suggested minimum, 1.5)
P = internal pressure, psi
A = cross-sectional area of the pipe based on outside diameter, in.^{2}
Δ = angle of deflection, degrees
F_s = pipe to soil friction based on half the circumference of the pipe, lb/ft

Restrained Length Design for Vertical Up Bends

Vertical up bends (shown in Figure 11-12) are assumed to act in the same manner as horizontal bends. The soil underneath a vertical up bend can be depended upon to provide bearing resistance.

Therefore, the design equation for a vertical up bend is the same as that for a horizontal bend (Eq 11-13) with both bearing and frictional resistance.

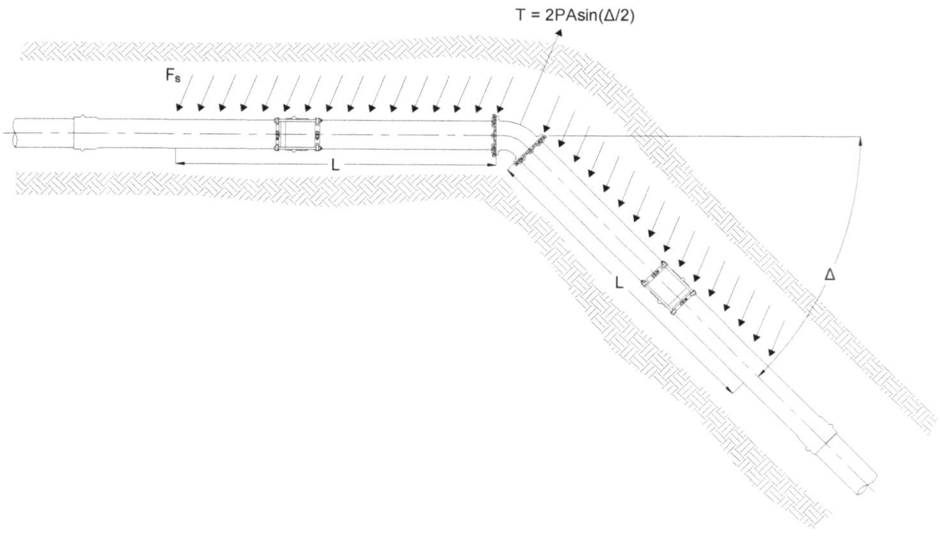

Figure 11-11 Restrained length design for a vertical down bend

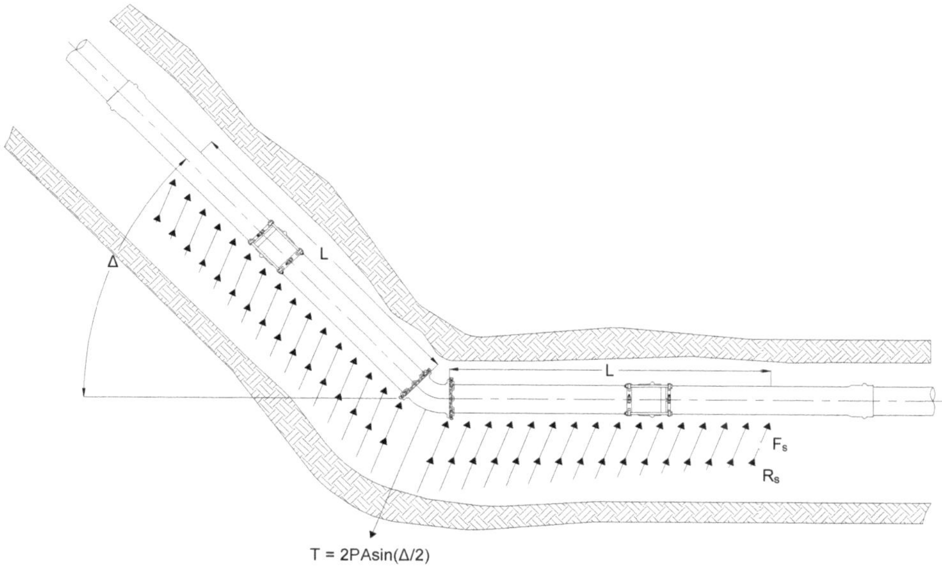

Figure 11-12 Restrained length design for a vertical up bend

Restrained Length Design: Other Considerations

The discussion to this point has involved fittings in isolation from one another. In a distribution system, it is common for several fittings to be in close proximity to one another. With thrust blocks, appropriate blocks are placed at each fitting. In a restrained system, the location of one fitting relative to another may mean that the restrained lengths for the two fittings overlap. This is known as encroaching restrained lengths.

Where the restrained lengths overlap between fittings, all of the joints between those fittings will be restrained. Therefore, no resultant thrust force will develop in that fully restrained section of pipe. It is important for the designer to determine which restrained lengths overlap and where the first unrestrained joint could be located in the installed pipeline. Since a thrust force will develop at that first unrestrained joint, the designer should determine if and how the thrust force may transfer to adjacent piping. That thrust force must be mitigated in some manner, either with additional restraint or with structural accommodation such as thrust blocks. Unfortunately, there is no consensus on a method to address encroaching restrained lengths.

A common fitting arrangement involving encroaching restrained lengths is a vertical offset. A vertical offset uses a vertical down bend and a vertical up bend to lower the elevation of a pipeline in order to avoid some kind of obstruction (such as a storm drain, other utilities, a creek bed, etc.). Generally, the distance between the upper bend and the lower bend is shorter than the restrained length requirement of the vertical down bend. As a result, the restrained lengths for the two fittings overlap and all joints between the two fittings are restrained. That means that the resultant thrust component in the oblique leg of the offset has now been eliminated. One way to address this would be to simply restrain back along the upper horizontal section for a distance determined by Eq 11-17 and along the lower horizontal section for a distance determined by Eq 11-12.

A very conservative approach to deal with an encroaching restrained length situation would be to treat the piping entering the first of a string of closely located fittings as if it is a dead end. Doing this eliminates all thrust force acting at that first fitting by transferring it via frictional resistance to the surrounding soil. Depending on other nearby fittings or other job specific conditions, it may be preferable to use a thrust block.

In some cases, when dealing with a dead-end thrust such as the termination of an existing pipeline, insertion of a valve, or the installation of an offset in an existing line, it is prohibitive to excavate and restrain the length of piping that would be required to satisfy the restrained length requirement of a dead end. In those cases, the designer can consider the use of a thrust collar in lieu of installing a large number of restraints. A thrust collar has a ring anchored to the pipe and a concrete thrust block poured around that ring and pipe and extends into the surrounding soil. The thrust collar should be sized in a manner similar to that required for a dead end.

Design Examples

The following design examples are provided to give the user a general idea of the scope and magnitude of the numbers involved in the design of a common PVC pipeline. The following example calculations will all be based on the following design criteria.

Pipe	AWWA C900 PVC pipe
Nominal pipe diameter	12 in.
Soil type	Sand and gravel with clay (thrust block calculations)
Soil type	CL (restrained length calculations)
Trench type	Type 5 (Chapter 7)
Depth of bury	4 ft
Test pressure	150 psi
Safety factor	1.5

For restrained length calculations, the manufacturers of joint restraining devices, among others, have published tabular values of R_s, F_s, and F_{sb}, or they can be calculated with the information provided earlier in this chapter. For the sake of space, those calculations are not included in these examples. Based on the stated design criteria, the values for the bearing and friction factors are:

R_s = 1,563.6 lb/ft
F_s = 295.8 lb/ft
F_{sb} = 425.4 lb/ft

For horizontal bends, the resultant thrust force at a 90 degree horizontal bend can be found using Eq 11-1.

$$T = 2PA\sin(\Delta/2)$$

where:

T = resultant thrust force, lb
P = internal pressure, 150 psi
A = cross-sectional area of the pipe based on outside diameter, 136.9 in.^{2}
Δ = angle of deflection, 90 degrees
T = 2(150)(136.9)sin(90/2)
T = 29,041 lb

The bearing area for a thrust block for this horizontal bend can be found using Eq 11-2 with a bearing strength of the soil taken from Table 11-1.

$$A = (T/q_u) \times SF$$

where:

A = bearing area of thrust block, ft^2
T = resultant thrust force, 29,041 lb
q_u = soil bearing capacity, 2000 lb/ft^2
SF = safety factor, 1.5
A = $(29{,}041/2000) \times 1.5$
A = 22.8 ft^2

Therefore, the bearing area of the thrust block should be approximately 3.5 ft high by 6.5 ft wide.

The same horizontal bend designed with restrained joints in lieu of thrust blocks can be found using Eq 11-13.

$$L = SF \times PA\tan(\Delta/2) / \left(F_s + \frac{1}{2}R_s\right)$$

where:

SF = safety factor, 1.5
P = internal pressure, 150 psi
A = cross-sectional area of the pipe based on outside diameter, 136.9 $in.^2$
Δ = angle of deflection, 90 degrees
F_s = pipe to soil friction based on half the circumference of the pipe, 295.8 lb/ft
R_s = bearing resistance of the soil along the pipe, 1563.6 lb/ft
L = $1.5 \times (150)(136.9)PA\tan(90/2)/(295.8 + ½ \times 1563.6)$

As a result, the design will require a total of four restraints—two for the joints of the fitting and two* for the pipe joints 20 ft away from the fitting.

REFERENCES

American Association of State Highway and Transportation Officials (AASHTO). T 223. 1967. Standard Method of Test for Field Vane Shear Test in Cohesive Soil. Washington, DC: American Association of State Highway and Transportation Officials.

ASTM International. 2012. D698. Standard Test Methods for Laboratory Compaction Characteristics of Soil Using Standard Effort (12 400 ft-lbf/ft^3 (600 kN-m/m^3)). West Conshohocken, Pa.: ASTM International.

ASTM International. 2018. D1586. Standard Test Method for Standard Penetration Test (SPT) and Split-Barrel Sampling of Soils. West Conshohocken, Pa.: ASTM International.

ASTM International. 2016. D2166. Standard Test Method for Unconfined Compressive Strength of Cohesive Soil. West Conshohocken, Pa.: ASTM International.

ASTM International. 2017. D2487. Standard Practice for Classification of Soils for Engineering Purposes (Unified Soil Classification System). West Conshohocken, Pa.: ASTM International.

ASTM International. 2015. D2573. Standard Test Method for Field Vane Shear Test in Saturated Fine-Grained Soils. West Conshohocken, Pa.: ASTM International.

Connections Bulletin. 2011. PD-6 Thrust Restraint Design Equations and Soil Parameters for Ductile Iron and PVC Pipe. Eastland, Tex.: EBAA Iron Sales, Inc.

Ductile Iron Pipe Research Association. 2016 (7th ed.). *Thrust Restraint Design for Ductile Iron Pipe.* Golden, Colo: Ductile Iron Pipe Research Association.

* Assuming standard length (20 ft) bell-and-spigot PVC pipe.

Kennedy, H. Jr.; Shumard, D. and Cary, M. 1989. Investigation of Pipe-to-Soil Friction and Its Effect on Thrust Restraint Design for PVC and Ductile Iron Pipe, EBAA Iron Inc. Presented at AWWA Distribution Systems Symposium, September 1989.

Kennedy, H. Jr.; Shumard, D. and Cary, M. 1990. *PVC Pipe Thrust Restraint Design Handbook*. Eastland, Tex.: EBAA Iron Inc.

Moser, A. and Folkman, S. 2008. *Buried Pipe Design.*

Potyondy, J.G. 1960. *Skin Friction Between Cohesive Granular Soils and Construction Materials,* Department of Civil Engineering, Technical University of Nova Scotia.

Chapter 12

Other Design Considerations

AXIAL DEFLECTION AND BENDING

PVC pipe's flexibility and/or axial deflection tolerance at joints enables it to be used for curved alignments and where other than straight alignment is required during installation. PVC pipe's resiliency also benefits from its bending ability (flexibility) and axial strain capacity. Ground movement or displacement may occur in response to various unanticipated conditions or unforeseen changes in conditions in the pipe-soil system, such as the following:

- Displacement associated with earthquakes
- Movement caused by tidal or groundwater conditions
- Seasonal variation in soil conditions caused by moisture content changes in expansive, organic soils and freeze-thaw cycles
- Movement resulting from excessive surface or traffic loading
- Uneven settlement of the pipe bedding
- Improper installation procedures: for example, nonuniform foundation, unstable bedding, or inadequate embedment consolidation

PVC pipe's ability to deform or bend and move away from external pressure concentrations without fracturing makes it highly resilient. The material's strain capacity and use of flexible joints enhances PVC pipe's ability to yield to external forces, thereby reducing risk of severe damage or failure.

When installing PVC pipe, gradual changes in alignment of the pipe may be accomplished without the use of elbows, sweeps, or other direction-change fittings. Controlled longitudinal bending of the pipe, within acceptable limits, can safely be accommodated by PVC pipe itself. Longitudinal movement of the pipeline can also be accommodated through limited axial joint deflection.

The safe allowable axial joint deflection depends upon the joint type, dimensions, and insertion depth. The print line on PVC pipe manufactured to ANSI/AWWA C900 includes the allowable axial joint deflection angle. Axial joint deflection limits should be obtained from the manufacturer for PVC pipe products not manufactured per ANSI/AWWA C900. Depending on pipe size and joint design, the allowable axial deflection for gasketed PVC pipe joints generally fall within the range of one-third of a degree (⅓°) up to three degrees (3°). The allowable axial joint deflection is dependent upon proper assembly. Proper assembly requires that the insertion mark, provided around the circumference near the pipe spigot end, remain visible near the bell-end of the adjoining pipe. The pipe spigot end should not be "homed" or fully inserted because when the leading edge of the spigot end makes contact with the tapered neck of the adjoining bell, the allowable deflection angle is reduced to zero. With segmented PVC pipe, some axial joint deflection tolerance can be important for handling of both horizontal and vertical offsets.

Higher deflection couplings are available for less gradual changes in alignment. Couplings axial deflection limits also vary and need to be established by the coupling manufacturer. Whenever substantial changes in alignment are needed, fittings should be used.

Allowable Bending Limits

Proper curved alignment design and installation requires that bending of PVC pipe be managed using conservative, PVC pipe-specific, criteria. The following discussion regarding allowable bending also applies to PVCO pipe.

Mathematical relationships for the longitudinal bending of pressurized tubes have been derived by Reissner. These relationships compare favorably to those of Timoshenko and others. The critical limit to bending of PVC pipe is long-term tensile stress. Axial bending also causes a very small amount of ovalization or diametric deflection of the pipe.

ANSI/AWWA C900 PVC pipe has short-term strengths of 7,000–8,000 psi (48.26–55.16 MPa) in tension and 11,000–15,000 psi (75.84–103.42 MPa) in flexure. The long-term strength of PVC pipe in either tension, compression, or flexure is conservatively assumed to equal PVC pressure pipes required minimum hydrostatic design basis (HDB) of 4,000 psi (27.58 MPa). HDB is explained and discussed in Chapter 5. Application of a 2.5:1 safety factor results in an allowable long-term tensile stress due to bending of 1,600 psi (11.03 MPa) at 73.4°F (23°C) and below.

The bending of PVC pressure pipe larger than nominal 6 in. (150 mm) is not recommended for segmented (elastomeric gasketed) pipe because of the higher forces required to bend larger pipe and associated difficulty in preventing excessive axial joint deflection. For PVC pipe with continuous joints, e.g., thermally fused or solvent cemented, there is no size limit for pipe bending.

The equation for allowable bending stress S_b is

$$S_b = (\text{HDB} - S_t)\frac{T'}{\text{SF}} \qquad \text{(Eq 12-1)}$$

where:

HDB = hydrostatic design basis of PVC pressure pipe, psi [4,000 psi (27.58 MPa)]
S_t = HDB/2 = tensile stress from longitudinal thrust, psi
T' = temperature rating factor (see Table 5-1)
SF = safety factor = 2.5

NOTE: The longitudinal stress at any internal pressure is only half as much as the hoop stress. The longitudinal stress from thermal expansion and contraction can generally be ignored in buried gasketed joint piping. Longitudinal thermal stresses should be considered whenever restrained joints will be used.

Using Eq 12-1, the maximum allowable bending stress (S_b) for restrained joint PVC pressure pipe is calculated as follows.

$$(S_b) \text{ at } 73.4°\text{F } (23°\text{C}) = \left[4{,}000 - \frac{4{,}000}{2}\right]\frac{1.0}{2.5} = 800 \text{ psi } (5.52 \text{ MPa})$$

The mathematical relationship between stress and moment induced by longitudinal bending of pipes is

$$M = \frac{S_b I}{c} \qquad \text{(Eq 12-2)}$$

where:

M = bending moment, in.-lb
S_b = allowable bending stress, psi
c = $D_o/2$ = distance from extreme fiber to neutral axis, in.
I = moment of inertia, in.^{4}

$$\left(\frac{\pi}{64}\right)\left(D_o^4 - D_i^4\right) = 0.049087\left(D_o^4 - D_i^4\right) \qquad \text{(Eq 12-3)}$$

where:

D_o = average outside diameter, in.
D_i = average inside diameter, in.
= $D_o - 2t_{avg}$, where:
t_{avg} = average wall thickness, in.
t_{avg} = t_{min} + 6% of t_{min}
t_{min} = minimum wall thickness, in.

Assuming that the bent length of pipe conforms to a circular arc after backfilling and installation (Figure 12-1, the longitudinal bending relationships for which are provided

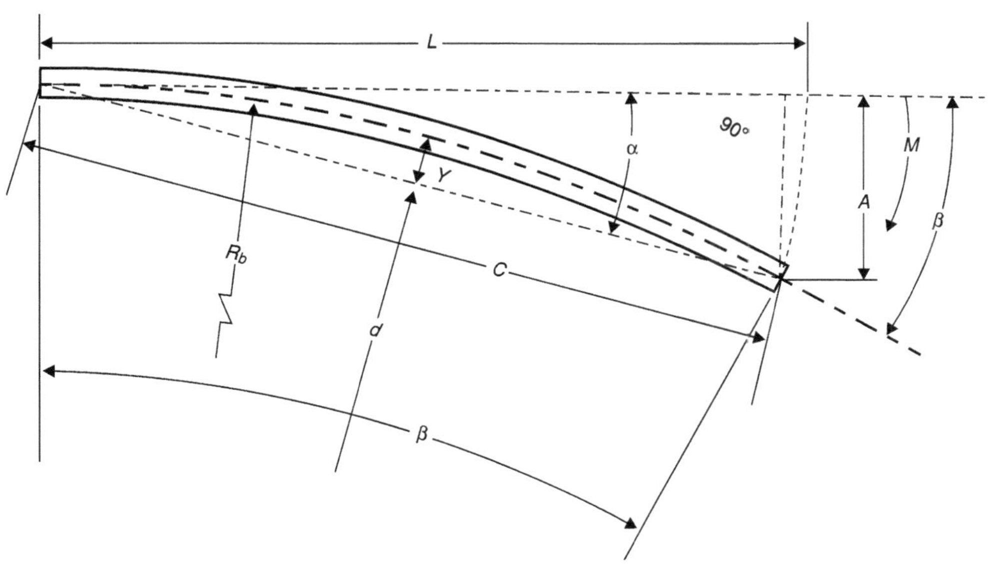

Figure 12-1 PVC pipe longitudinal bending diagram

Table 12-1 Longitudinal bending relationships for Figure 12-1

$\alpha = \beta/2$	$L = \dfrac{R_b \pi}{90}\alpha$
$Y = R_b - d$	$C = 2R_b \sin\beta/2 \approx L$
$\beta = \dfrac{360L}{2\pi R_b}$	$d = R_b \cos\beta/2$
$A = 2R_b(\sin^2 \beta/2) = C \sin\beta/2 = L \tan\alpha$	$M = \dfrac{S_b I}{c}$

in Table 12-1), the minimum radius R_b of the bending circle can be found by Timoshenko's equation:

$$R_b = \frac{EI}{M} \qquad \text{(Eq 12-4)}$$

Combining Eq 12-2 and Eq 12-4 gives

$$R_b = \frac{ED_o}{2S_b} \qquad \text{(Eq 12-5)}$$

The central angle β subtended by the length of pipe is

$$\beta = \frac{360L}{2\pi R_b} = \frac{57.30L}{R_b} \qquad \text{(Eq 12-6)}$$

where:

R_b = minimum bending radius, in.
E = modulus of tensile elasticity, psi
M = bending moment, in.-lb
S_b = allowable bending stress, psi
L = pipe length, in.

L and R_b are both in the same units, and the angle of lateral offset (α) of the curved pipe from a tangent to the circle is degrees

$$\alpha = \frac{\beta}{2}, \text{ degrees} \qquad \text{(Eq 12-7)}$$

The offset A at the end of the pipe from the tangent to the circle can then be calculated as

$$A = 2R_b(\sin^2 \beta/2) = 2R_b(\sin\alpha)^2 \qquad \text{(Eq 12-8)}$$

where:

A = offset at the end of the pipe from the tangent to the circle, in.

Assuming that during installation the pipe is temporarily fixed at one end and acts as a cantilevered beam, then the lateral force required at the free end to achieve the offset *A* may be determined by the following equation:

$$P = \frac{3EIA}{L^3} \qquad \text{(Eq 12-9)}$$

where:

P = lateral offset force, lb
E = modulus of tensile elasticity, psi
I = moment of inertia, in.^{4}
A = offset at free end, in.
L = pipe length, in.

The mathematical relationship between the bending offset angle α, the offset A, the lateral offset force P, and the minimum bending radius R_b is shown in Figure 12-1. The bending limits calculated by the equations listed above are without allowance for joint offset and without consideration of the stresses imposed upon the joint. Because of the characteristics of a particular joint design, it is possible that a manufacturer's recommended bending radius may be greater than those presented in Table 12-2. The 800 psi, maximum allowable bending stress (S_b) for restrained joint PVC pressure pipe, occurs at a bending radius (R_b) of about 250 D_o (R_b = 250 D_o).

Longitudinal bending of gasket-joined pipe larger than nominal size 6-in. (150-mm) diameter is not recommended because excessive joint deflection and stress are very difficult to control under high bending force conditions.

Table 12-2 Allowable longitudinal bending for fusion-joined PVC pressure pipe

Nominal Size *in. (mm)*	Minimum Bending Radius (R_b) CIOD* *ft (m)*	Minimum Bending Radius (R_b) IPS† *ft (m)*
4 (100)	100 (30.5)	
6 (150)	144 (43.9)	
8 (200)	189 (57.6)	
10 (250)	231 (70.4)	
12 (300)	275 (83.8)	
14 (350)	319 (97.2)	292 (89.0)
16 (400)	363 (110.6)	333 (101.5)
18 (450)	406 (123.7)	375 (114.3)
20 (500)	450 (137.2)	417 (127.1)
24 (600)	538 (164.0)	500 (152.4)
30 (750)	667 (203.3)	620 (189.0)
36 (900)	798 (243.2)	742 (226.2)
42 (1,050)	927 (282.6)	
48 (1,200)	1,058 (322.5)	

Source: ANSI/AWWA C605-13
Underground Installation of Polyvinyl Chloride (Pvc) Pressure Pipe and Fittings for Water.
* Cast iron outside diameter: Outside diameters that are equivalent to iron pipes and fittings, in accordance with the ductile-iron or cast-iron (CIOD) sizing system (ANSI/AWWA C110/A21.10).
† Iron pipe size: Outside diameters that are equivalent to steel pipes and fittings, in accordance with the steel (IPS) sizing system (ASME B36.10M).

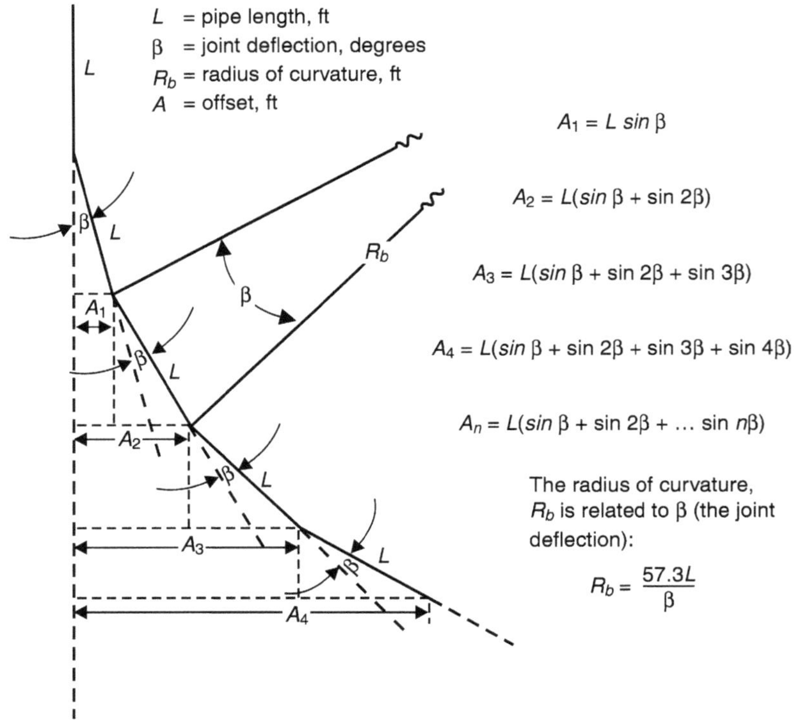

Figure 12-2 PVC pipe joint offset

Joint Offset

When a desired change of direction in a PVC pipeline exceeds the permissible joint offset angle β for a given length of pipe, the total offset required can be distributed through several pipe lengths (Figure 12-2). Calculation of required distribution of total offset in PVC pipe is demonstrated in the following example.

- Pipeline using ANSI/AWWA C900 8-in. PVC DR 18 pipe in 20-ft lengths.
- Desired change of direction is 10°.
- End offset of one 20-ft length = A1 (Figure 12-2).

  β = 3.0° per pipe length, maximum end offset
  (Four lengths of 8-in. × 20-ft pipe at 2.5° are required.)

Resultant total offset for the pipeline over four pipe lengths:

$$\begin{aligned} A_4 &= 20[\sin 2.5° + \sin(2 \times 2.5°) + \sin(3 \times 2.5°) + \sin(4 \times 2.5°)] \\ &= 20\,(0.0436 + 0.0872 + 0.1305 + 0.1736) \\ &= 20\,(0.4349) \\ &= 8.7 \text{ ft} \end{aligned}$$

Performance Limits in Longitudinal Bending

Bending strain. Longitudinal bending strain ε and longitudinal bending stress S_b for PVC pipe at different bending radius ratios are tabulated in Table 12-3 using Eq 12-10:

$$\varepsilon = S_b/E = D_o/2R_b \qquad \text{(Eq 12-10)}$$

Bending ovalization (diametric or ring deflection). As a thin tube is bent longitudinally, it will change into an approximate elliptical shape. This effect has been ignored as

Table 12-3 Longitudinal bending stress and strain in PVC pipe at 73.4°F (23°C)

Bending Radius Ratio, R_b/D_o	Elastic Modulus, E psi	Elastic Modulus, E (MPa)	Bending Strain, ε in./in.	Bending Stress, S_b psi	Bending Stress, S_b (MPa)
25	400,000	(2,758)	0.0200	8,000	(55.16)
50	400,000	(2,758)	0.0100	4,000	(27.58)
100	400,000	(2,758)	0.0050	2,000	(13.79)
200	400,000	(2,758)	0.0025	1,000	(6.89)
250	400,000	(2,758)	0.0020	800	(5.52)
300	400,000	(2,758)	0.0017	667	(4.60)
500	400,000	(2,758)	0.0010	400	(2.76)

Note: These values also apply to PVCO pipe.

insignificant in previous calculations on longitudinal bending. Ring deflection is usually expressed as

$$\text{Deflection} = \delta = \frac{\Delta Y}{D} \quad \text{(Eq 12-11)}$$

$$\%\ \text{Deflection} = 100\delta = 100\frac{\Delta Y}{D} \quad \text{(Eq 12-12)}$$

where:

ΔY = reduction in diameter, in.
D = diameter, in.

The mathematical relationships for thin pressurized tubes between ring deflection and axial bending, as derived by E. Reissner, are as follows:

$$\delta = \frac{\Delta Y}{D} = -\left(A_1\alpha^2\right)\left[\left(\frac{2}{3}\right) + \frac{71 + 4\lambda}{135 + 9\lambda}\left(A_1\alpha^2\right)\right] \quad \text{(Eq 12-13)}$$

with λ and $(A_1\alpha^2)$ defined as

$$\lambda = \frac{12(1 - u^2)PD_m^{\ 3}}{8Et^3} \quad \text{(Eq 12-14)}$$

$$\left(A_1\alpha^2\right) = \frac{1}{6}\left[\frac{18(1 - u^2)}{12 + 4\lambda}\right]\frac{D_m^{\ 4}}{R^2t^2} \quad \text{(Eq 12-15)}$$

where:

D_m = mean pipe diameter, in.
u = Poisson's ratio (0.38 for PVC)
P = internal pipe pressure, psig
E = modulus of elasticity, psi
t = pipe thickness, in. (use $t_{nom} = 1.06 \times t$)
R = bending radius of pipe, in.

Example: Calculate the percent ring deflection that results from bending a 4-in., DR 14 PVC pressure pipe to a bending radius that is 300 times the outside diameter and pressurized to 100 psi:

$$\lambda = \frac{12(1 - 0.38^2)100(4.800 - 0.364)^3}{8(400{,}000)(0.364)^3}$$

$$= \frac{1{,}200(0.8556)(87.292)}{(3{,}200{,}000)(0.04823)} = 0.581$$

$$(A_1\alpha^2) = \frac{1}{16}\left[\frac{18(1 - 0.38^2)}{12 + 4(0.581)}\right]\frac{(4.800 - 0.364)^4}{(1{,}200)^2(0.364)^2}$$

$$= \frac{1}{16}\left[\frac{18 \times 0.8556)}{12 + 2.324}\right]\frac{387.2275}{(1{,}440{,}000)(0.1325)}$$

$$= 0.000136$$

$$\delta = -(0.000136)\left[\left(\frac{2}{3}\right) + \frac{71 + 4(0.581)}{135 + 9(0.581)} \times 0.000136\right]$$

$$= -(0.000136)\left[\left(\frac{2}{3}\right) + 0.0000711\right]$$

$$= -0.0000907$$

$$= -0.009\%$$

A negative value indicates that the deflection deforms the pipe such that the diameter aligned with the plane of bending is decreased by 0.009% while the diameter perpendicular to the plane of bending is increased by the same amount.

From the preceding example, for a bending radius of 300 times the pipe diameter, a close approximation of deflection can be calculated with the following equation:

$$\delta = \frac{\Delta Y}{D_m} = -\frac{2}{3}(A_1\alpha^2) = -\frac{(1 - u^2)(D_m)^4}{16R^2t^2} \qquad \text{(Eq 12-16)}$$

Analysis of similar examples has shown that the amount of deflection resulting from bending is negligible. At pipe bending radii ≥ 250 times the pipe diameter, the ring deflection from bending will be less than 0.1 percent.

EXPANSION AND CONTRACTION

All pipes expand and contract with changes in temperature. The linear expansion and contraction of any pipe is dictated by the pipe material's coefficient of thermal expansion together with the magnitude of the temperature change (ΔT). Pipe expansion and contraction is independent of diameter and wall thickness.

Table 12-4 PVC coefficient of thermal expansion in relation to other pipe materials

Piping Material	Coefficient *in./in./°F*	Expansion *in./100 ft/10°F*
PVC	3.0×10^{-5}	0.36
HDPE	1.2×10^{-4}	1.44
ABS	5.5×10^{-5}	0.66
Asbestos cement	4.5×10^{-6}	0.05
Aluminum	1.3×10^{-5}	0.16
Cast iron	5.8×10^{-6}	0.07
Ductile iron	5.8×10^{-6}	0.07
Steel	6.5×10^{-6}	0.08
Clay	3.4×10^{-6}	0.04
Concrete	5.5×10^{-6}	0.07
Copper	9.8×10^{-6}	0.12

Table 12-5 Length variation per 10°F (5.6°C) ΔT for PVC pipe

Pipe Length		Length Change	
ft	*(m)*	*in.*	*(mm)*
20	(6.1)	0.072	(1.83)
10	(3.0)	0.036	(0.91)

Coefficients of Thermal Expansion

Approximate coefficients of thermal expansion for different pipe materials are presented in Table 12-4. Expansion and contraction of pipe in response to change in temperature can vary slightly due to changes in the composition of the pipe material. However, the coefficients shown in Table 12-4 are representative for the listed pipe materials. Table 12-5 displays expansion and contraction change per length of PVC pipe. PVC pipe length variation caused by temperature change is shown graphically in Figure 12-3.

A good general rule in design of PVC piping systems is to allow ⅜ in. (10 mm) of length variation for every 100 ft (30 m) of pipe for each 10°F (5.6°C) change in temperature. The coefficient of thermal expansion is the same for PVCO and PVC pipe materials.

PVC pipe with gasketed joints, if properly installed (i.e., with pipe spigot inserted into bell joints up to manufacturer's insertion mark), will accommodate substantial thermal expansion and contraction. If gasketed joints are used, thermal expansion and contraction is not a significant factor in system design within the accepted range of operating temperatures for buried PVC water pipe.

For systems where expansion and contraction cannot be accommodated by gasketed joints, the stresses resulting from extreme temperature changes should be considered using the following equation:

$$S' = EC_T(t_1 - t_o) \quad \text{(Eq 12-17)}$$

where:

S' = stress, psi
E = modulus of tensile elasticity, psi
C_T = coefficient of thermal expansion, in./in./°F
t_1 = highest temperature, °F
t_o = lowest temperature, °F

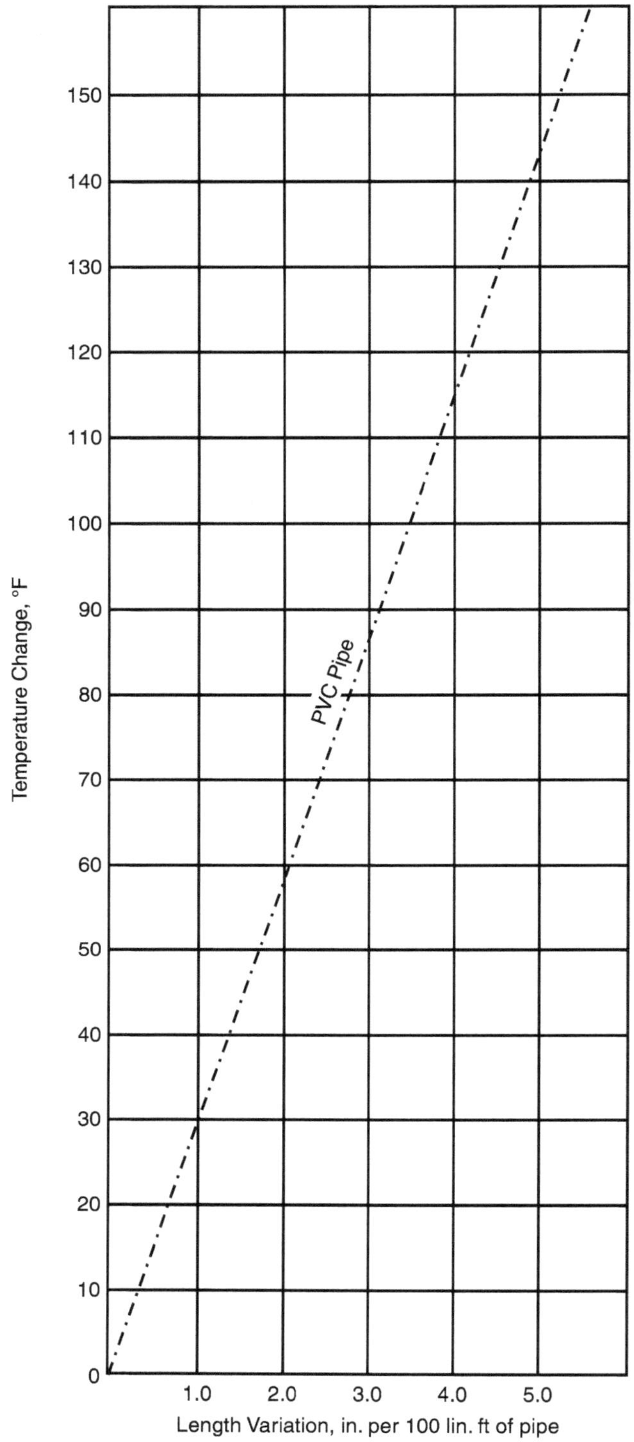

Figure 12-3 Length variation of unrestrained PVC pipe as a result of temperature change

Example: Calculate the stress resulting from a temperature change from 120°F to 30°F in a restrained PVC pipe.

$S' = (400{,}000) \times (3.0 \times 10^{-5}) \times (120 - 30)$

$S' = 1{,}080$ psi

This is below the minimum short-term tensile strength of 7,000 psi. The 7,000 psi value may be used as a conservative minimum value for the short-term compressive strength.

METHODS FOR LOCATING BURIED PVC PIPELINES

The best method for locating underground pipelines is the use of accurate "as-built" drawings. Utilities and municipalities retain up-to-date records of installations, but may encounter problems if such records do not accurately reflect field work or when landmarks (curbs, hydrants, light standards) have been moved, road grades have been altered, etc. In anticipation of these circumstances, alternate methods of locating nonconductive buried PVC pipelines may be required.

Tracer wire or tape. An inexpensive metallic tracer wire can be buried with the pipe. Plastic-coated metal tapes or strips designed specifically for this purpose are also available. These brightly colored strips can also serve as warning markers. Different colored tracer tapes may be used to more specifically identify various underground utilities. The utility or owner should be contacted to determine where, in relation to the pipe, the tracer wire or tape should be installed.

An electronic transmitter can be used to induce a signal onto a metallic tracer wire or plastic-coated metallic tape from the ground surface. Detection of pipelines at greater depths can be successfully achieved through conductive detection where the detector is physically connected to the tracer wire or tape. These conductors can be connected to a valve box or hydrant where they can be accessed.

Acoustical systems. Acoustical methods for locating PVC pressure pipe have proven to be quite accurate. Acoustical locating systems, comprised of a transmitter and a receiver, are commercially available. Transmitters introduce a sound wave into the system, which propagates through the PVC pipe. The receiver is used to locate the sound and thus locate the line. Transmitters can be attached directly to lines or more easily attached to connected fire hydrants, service lines or water meters.

Ground penetrating radar. Alternatively, electronic pipe detection devices called "ground-penetrating radar" may be used. These systems rely on pulses of energy (similar to seismographs) which, when directed downward, send back echoes. Trained operators can identify the depth and location of buried pipes using such devices.

REFERENCES

Plastics Pipe Institute. 2001. Thermal Expansion and Contraction of Plastic Pipe, PPI Technical Report, PPI-TR-21. Irving, Tex.: Plastics Pipe Institute.

Reissner, E. 1959. On finite bending of pressurized tubes. *Journal of Applied Mechanics Transactions of ASME.*

Timoshenko, S.P. and Young, D.H. 1962 (4th ed.). *Elements of Strength of Materials.* Van Nostrand Co., Princeton, NJ.

Uni-Bell PVC Pipe Association. 2012 (5th ed.). *Uni-Bell Handbook of PVC Pipe—Design and Construction.* Dallas, Tex.: Uni-Bell PVC Pipe Association.

This page intentionally blank.

Appendix **A**

Requirements for Establishing the Hydrostatic Design Basis (HDB) of PVC Compounds for Pipe or Fittings

Foreword

This appendix to AWWA Manual 23 describes the requirements for establishing the Hydrostatic Design Basis (HDB) ratings for the compounds used to make PVC pipe and fittings. These requirements are derived from over 50 years of progressive development by the HSB (Hydrostatic Stress Board) of PPI* (Plastics Pipe Institute, Inc.). PPI is a trade association of plastic pipe manufacturers.

The requirements presented herein are for compounds that are PVC resin based. The manufacturer is responsible for the production of pipe or fittings that provide long-term strength consistent with these requirements. The purchaser may require an affidavit

* All tables in Appendix A are adapted from Plastics Pipe Institute, Inc.

from the manufacturer, supplier, or a third-party that a PVC compound complies with the applicable requirements of this Appendix.

SECTION A1: REQUIREMENTS FOR HDB RATING OF PVC COMPOUNDS

Sec. A1.1 HDB Rating Data and Evaluation Requirements

A1.1.1 HDB rating. An HDB rating shall be assigned to compounds that comply with the full stress rupture data requirements of "Standard Test Method for Obtaining Hydrostatic Design Basis for Thermoplastic Pipe Materials," ASTM D 2837, for at least one pipe lot and the minimum requirements of the I-2 level of the Initial HDB rating for each of at least two additional lots. At least one of these three lots of pipe shall be made by a pipe producer on commercial production equipment.

A1.1.2 Initial HDB rating. An Initial HDB rating is for a limited duration and shall be assigned to those compounds that do not yet comply with the full requirements of Section A1.1.1, but satisfy the applicable minimum preliminary data requirements that are detailed in the Sec. A1.1.2.1

A1.1.2.1 Test durations and data requirements for initial ratings. Initial HDB ratings must be advanced through specified stress rupture data levels until the full data requirements of the HDB rating are satisfied. Failure to advance an Initial HDB rating will cause the compound's HDB rating to expire. The Initial HDB rating shall meet the requirements of the latest revision of "Standard Test Method for Obtaining Hydrostatic Design Basis for Thermoplastic Pipe Materials," ASTM D 2837, except that the submitted test data need only meet the requirements given in Tables A-1 and A-2:

Table A-1 Minimum data requirements for initial ratings

Initial Level	Time of Data Submission	Minimum Data Required	
		Lots	At Level
I-2	Initial	One	I-2
I-4	Initial or within 6 months after Level 2	One	I-4
I-6	Initial or within 6 months after Level 4	One	I-6
I-8	Initial or within 6 months after Level 6	One	I-8

Table A-2 Minimum test duration and data distribution requirements for initial levels

Initial Level	Requires Data for at Least (hr)	Total number of data points	Number of Data Points					
			Less than 1000 (hr)	Over 2000 (hr)	Over 4000 (hr)	1000–6000 (hr)	Over 6000 (hr)	Over 8000 (hr)
I-2	2,000	10	6	1	—	—	—	—
I-4	4,000	10	6	1	1	—	—	—
I-6	6,000	12	6	—	—	3	1	—
I-8	8,000	15	6	—	—	3	1	1
I-10 (final)	10,000	18	(Refer to latest issue of ASTM D 2837 for data distribution requirements)					

The requirements of Section A1.1.1 shall be met within six months after those of I-8 initial level are met.

Note: It is generally recognized that the equipment and conditions employed in the mixing and processing of an HDB rated compound will influence the compound's long-term ability to handle stress. As an example, the same pipe formulation that in one case is premixed prior to pipe extrusion, but in another depends entirely on the extrusion process for proper dispersion of its ingredients, may not yield the same long-term strength under both conditions. Accordingly, it is the manufacturer's responsibility to ensure equivalence of the processing conditions under which the HDB rating is established.

Sec. A1.2 Optional Data and Evaluation Requirements for HDB Rating

As an option to the testing requirements in Sec. A1.1, the following procedure may be used in evaluating the stress-rupture data for (1) second and third lots of pipe required in Sec. A1.1.1 to obtain a final HDB rating, (2) lots of pipe used to demonstrate the effects of substitution of "apparently identical ingredients" as described in Sec. A1.5, and (3) lots of pipe used to show the effects of changes in the level of compounding ingredients (excluding stabilizers) up to ±50 percent from the levels contained in a compound with a final HDB rating.

A1.2.1 Optional data and analysis procedure. After five or more stress-rupture data points are developed, compute the long-term hydrostatic strength (LTHS), the 95 percent upper (UCL) and lower (LCL) confidence levels, and the lower confidence level ratio (R_{LCL} = LCL/LTHS) per ASTM D2837. Check these results against the following requirements:

(1) If $R_{LCL} \geq 0.85$, evaluate data in accordance with (2), (3), or (4) of this Section [A1.2.1]. If $R_{LCL} < 0.85$, incorporate additional data points as they become available and re-compute until $R_{LCL} \geq 0.85$.

(2) If the 100,000-hr LCL stress value is greater than 4,000 psi, this lot will always exceed the 3,830-psi long-term hydrostatic strength requirement for a 4,000-psi HDB rating. No further stress rupture testing is required to fulfill the requirements of Sec. A1.5 for apparently identical ingredients.

(3) If the 100,000-hr UCL stress value is less than 3,830 psi, this lot will not attain the required long-term hydrostatic strength to qualify for an HDB rating of 4,000 psi. The tests should be stopped unless an HDB rating below 4,000 psi is acceptable.

(4) If the 100,000-hr UCL stress value is more than 3,830 psi, and the 100,000-hr LCL stress value is less than 3,830 psi, the stress rupture testing should be continued to obtain more data points. Additional data points shall be included and evaluated until the lot qualifies or fails either under this method or Sec. A1.1.

Note: As an approximation, the UCL may be estimated from the LCL by the following equation:

$$\text{Log } UCL = \log LTHS + (\log LTHS - \log LCL)$$

Sec. A1.3 Resin Substitution Requirements

An alternate resin may be used in a HDB rated compound provided that all of the following conditions are met. First, the alternate resin has been used in at least one HDB rated compound having the same HDB rating as the subject compound. Second, the compound's ASTM D 1784, "Standard Specification for Rigid Poly(Vinyl Chloride) (PVC) Compounds and Chlorinated Poly(Vinyl Chloride) (CPVC) Compounds," cell classification must remain the same with the alternative resin as it was for the HDB rated compound. And third, the alternative resin must satisfy all of the substitution resin property requirements shown in Table A-3.

Table A-3 Substitution resin property requirements

Property	Test Method	Requirement
Type of material	—	PVC Homo-polymer
Inherent viscosity	ASTM D1243	0.88–0.96
Heat loss (water)	1 hr @ 221°F (105°C)	≤0.5 % by weight
Apparent bulk density	ASTM D1895	0.46–0.62 gm/cc
Compacted bulk density	—	0.54–0.72 gm/cc
RVCM	—	<3.2 ppm

RVCM: residual vinyl chloride monomer

A1.3.1 Initial HDB rating. The compound with the substitute resin shall be eligible for an Initial HDB rating upon the completion of hydrostatic rupture tests at 73°F (23°C), of pipe from one or more extrusion lots and in accordance with ASTM D2837, that consist of at least nine data points covering a range of failure times, in hours, spanning at least three log cycles (i.e., from 0.05 to 50 hr, or from 1 to 1,000 hr).

A1.3.2 HDB rating. A final HDB rating for the substitute resin compound must be completed within six months of the Initial HDB rating and requires at least six additional data points representing not less than three different extrusion lots. The failure time for these points shall be not less than 10 hr with at least two points over 2,000 hr. These additional data may be evaluated either in combination with, or separately from, those supplied for the initial HDB rating.

Note: The resin requirements given in this section are not specifications or standards that describe the requirements for all the PVC resins suitable for HDB rated compounds.

Sec. A1.4 Formulation Variability Limits

Within the limits given in Table A-4, the content of one or more ingredients may be varied without changing the HDB rating provided each adjusted ingredient meets the property requirements of Section A1 and the resultant formulation can be processed into pipe of acceptable quality:

A1.4.1 Higher variation limits. To establish a higher variation range, maximum and minimum levels for one or more ingredients, testing shall be performed as required in Sec. A1.1 for the compound when formulated with all ingredients (all additives except resin) at the identified maximum level and when formulated with all ingredients at their identified minimum level. Stress-rupture testing on one lot of pipe made from the maximum ingredient level formulation must meet the full requirements of ASTM D 2837. I-2 level (see Sec. A1.1.2.1) stress-rupture testing shall be performed on at least one lot of pipe made from the minimum ingredient level formulation. In addition, I-2 level stress rupture testing shall be performed on one other lot of pipe extruded on commercial production equipment that is made from a formulation lying within the identified maximum/minimum range. All three pipe lots must qualify for the original HDB rating.

A1.4.2 Percentage changes between two HDB rated compounds. The percent of a specific ingredient may be varied without further testing over the range between the percentages given in at least two compounds with the same HDB rating that differ only in the percent of that ingredient.

A1.4.3 Measurement variability. Because of the differing precisions of measuring devices and the normal variations inherent even in good handling techniques and processing equipment, the small variations in the amount of an ingredient in a compound shall not be considered changes in the formulation provided: (1) the norm is set at the percentage

Table A-4 Compound ingredient variation limits

Ingredient	Allowable Variation from that of the HDB Rated Compound
Titanium dioxide	±20%
Calcium stearate	±10%
Calcium carbonate	±20%, but not >5.0 phr or the original specified amount if greater than 5.0 phr
Paraffin wax	±10%
Polyethylene wax	±20%, but not >0.30 phr or the original specified amount if greater than 0.30 phr
Colorant	±20%
Process aid	±20%
Stabilizer	±20%, but not >±0.2 phr

given in the formulation and (2) variations over a reasonable number of batches are small and about equal on the average on the plus and minus side of the norm.

Sec. A1.5 Substitution Limits and Requirements for Apparently Identical Ingredients

For any ingredient that is ≤5 parts per hundred parts (by weight) of resin, substitution of an apparently identical ingredient from a different supplier with no change in amount shall be permitted when the least squares regression line for the substituted compound obtained with I-2 level test data produces: (1) a 100,000-hr hydrostatic strength that is not less than required to give the same hydrostatic design basis as that for the HDB rated compound and (2) a 50-year strength value that is not less than 85 percent of the 50-year strength value of the HDB rated compound.

Sec. A1.6 Substitution Limits and Requirements for Thermal Stabilizers

A new, or modified, stabilizer may be substituted for the original stabilizer provided that: (1) the substitute stabilizer is also already used in at least one HDB rated compound; (2) the cell classification, when determined in accordance with ASTM D 1784, is the same for the compound made with the substitute stabilizer as that of the original HDB rated compound; (3) the stabilizer level in the compound must be within the range of 0.3–1.0 parts per hundred parts of resin and the amount cannot be varied more than allowed by Sec. A1.3; and (4) the least squares regression line for pipe made from the stabilizer substituted compound obtained with I-2 level test data produces a 100,000-hr hydrostatic strength that is not less than required to give the same hydrostatic design basis as that for the HDB rated compound and a 50-year strength value that is not less than 85 percent of the 50-year strength value of the HDB rated compound. The tested pipe shall be manufactured using commercial production equipment.

Sec. A1.7 Substitution Limits and Requirements for Colorants

The amount of colorant in an HDB rated compound may be changed by as much as 0.5 PHR* without additional hydrostatic strength data, provided the colorant is neither a liquid nor includes a liquid component at ambient temperature conditions. Any type inorganic colorant that is used in the HDB rated compound may be substituted in part or

* PHR is the unit weight of an added ingredient per one hundred units of a compound's base resin.

in whole, by any other inorganic colorant provided the substituted colorant is limited to a maximum of 0.5 PHR. Organic colorants that are used in the HDB rated compound may be substituted in part or in whole, by any other organic colorant provided the substituted colorant is limited to a maximum of 0.5 PHR and that the substituted organic colorant is equivalent to the original organic colorant. Colorants are considered equivalent only if they are of the same chemical nature, same particle size classification, exhibit the same distribution or mixing properties, exhibit the same non-reactive properties with the base resin, and remain solid over the intended application range. A compound with a colorant change does not become a new base compound to which additional formulation changes can be made.

A1.7.1 Compounds with an HDB rating that has been established based on a specific range of colorant content. Whenever hydrostatic stress rupture data have been generated that qualify a compound at two different colorant levels, and thereby established a colorant range, then no additional hydrostatic pressure testing is required for variations in the amount of colorant within that established range.

Note: The manufacturer should make adequate tests to assure that the colorant changes do not adversely affect performance properties other than HDB rating, such as aging and weather resistance.

Sec. A1.8 Requirements for Molding Compounds

Compounds intended only for molded fittings and parts shall be evaluated in accordance with Secs. A1.8.1 and A1.8.2. Compounds intended for both extrusion and molding shall be evaluated in accordance with Secs. A1.1 – A1.7.

A1.8.1 Test specimens. The stress rupture test specimens shall be injection-molded tubes with weld lines that result from either a side gate or an end gate. The tube wall of each specimen shall be as uniform in thickness as is technically possible. The molding conditions used to mold the test specimens shall be consistent with molding conditions used for that material in actual end-use applications. The test specimen length between end enclosures shall be at least five times the specimen outside diameter.

A1.8.2 Testing requirements. The specimens shall be tested in the same manner as extruded pipe specimens. The data and evaluation requirements shall be the same as those required in Secs. A1.1 – A1.7 for the HDB rating of extruded pipe compounds except that the report shall include complete information about the specimen molding conditions and gating design.

SECTION A2 INGREDIENT QUALIFICATION REQUIREMENTS AND RANGE LIMITS FOR HDB RATED COMPOUNDS EXEMPT FROM FURTHER STRESS RUPTURE TESTING

Sec. A2.1 General Qualification Requirements and Compound Formulation Limits

ASTM D1784 cell class 12454 compounds formulated within the allowable range limits given in Table A-5 and comprised of ingredients or combinations of ingredients that meet the physical property requirements of Sec 2.2 shall not require further stress rupture testing to qualify for an HDB rating of 4,000 psi at 73°F (23°C).

Note: The pipe manufacturer should perform some stress-rupture testing on new combinations of ingredients to verify that the HDB is realized under the selected processing conditions. Proper HDB may be verified by use of either the Accelerated Regression Test as described in ASTM D 2241 or other long-term stress-rupture data generated in accordance with ASTM D1598 and analyzed by the method of least squares as described in ASTM D2837.

Table A-5 Generic compound ingredient requirements and range limits

Ingredient	Qualification Compliance	Allowable Range Limit (in parts per hundred parts of resin)
PVC resin	With Sec A2.2.1	100
Heat stabilizer	With Sec A1.6	0.3–1.0
Calcium stearate	With Sec A2.2.2	0.4–1.5
Paraffin wax	With Sec A2.2.3	0.6–1.5
Polyethylene wax	With Sec A2.2.4	≤0.3
Titanium dioxide	With Sec A2.2.5	0.5–3.0
Calcium carbonate	With Sec A2.2.6	≤5.0
Process aid	With Sec A2.3	≤2.0
Colorant	With Sec A1.7	≤2.0

Sec. A2.2 Individual Ingredient Qualification Requirements

The interchangeability of compound ingredients within the "range formulation" limits given in Sec A2.1 requires that the individual ingredients comply with the qualification requirements in this Section.

A2.2.1 PVC resin. Only suspension grade resins shall be permitted. The resin shall meet the property requirements given in Table A-3.

In addition, to qualify the resin's use in the "range formulation," an HDB rating of 4,000 psi at 73°F (23°C), stress-rupture testing of pipe extruded on commercial equipment from a PVC compound with components falling within the ranges given in Table A-5, but with liquid stabilizer and calcium carbonate at the maximum levels, shall be required. The test data must be obtained and evaluated according to Sec A1.1.

An I-2 level rating of 4,000, psi per Sec A1.1 shall be required on a commercial lot of pipe to qualify equivalent resin manufactured at a different plant.

A2.2.2 Calcium stearate. Commercial calcium stearates may be substituted one for another without additional pipe stress rupture testing provided: (1) the substitution is for the same amount, (2) the substitution in the compound is 1.5 parts or less per hundred parts of resin, and (3) the properties fall within the ranges given in Table A-6 or the ingredient has been qualified as functionally equivalent.

Calcium stearates not meeting the requirements in Table A-6 can qualify as a functionally equivalent ingredient upon successful completion of Sec. A2.3 testing requirements.

Note: The above requirements are not specifications or standards for calcium stearates. They do not describe all the commercial calcium stearates that are suitable for use in PVC pipe compounds. These requirements only serve to define members within a group of commercial calcium stearates that are apparently chemically identical and that have been found by physical/chemical analysis, by pipe testing, and by use of pipe in the field to be sufficiently similar in nature to produce pipe and fittings that exhibit basically the same properties and performance.

A2.2.3 Paraffinic hydrocarbon wax. Commercial paraffinic hydrocarbon waxes may be substituted one for another without additional stress rupture testing provided: (1) the substitution is for the same amount, (2) the substitution in the compound is 1.5 parts or less per hundred parts of resin, (3) the particle size is approximately the same, and (4) the properties fall within the ranges given in Table A-7 or the ingredient has been qualified as functionally equivalent.

Table A-6 Calcium stearate property requirements

Property*	Requirements
Melting point	293–329°F (145–165°C)
Chlorides, as chloride ion	≤0.3%
Free fatty acid	≤1.0%
Acid number†	200 ± 15
Iodine value†	≤5.0
Volatile content, at 221°F (105°C)	≤3.5%
Ash, as CaO, at 1832°F (1000°C)	9.00–12.0%
Iron	≤100 ppm
Color	White to off-white
Particle size	At least 95% through U.S Standard 20 mesh

* The test methods shall be those commonly used in the calcium stearate industry.
† Acid number and iodine value apply to the stearic acid raw material.

Table A-7 Paraffinic hydrocarbon wax property requirements

Property	Test Method*	Requirement
Chemical type		Hydrocarbon wax†
Congealing point	ASTM D938	149–169°F (65–76°C)
Viscosity at 210°F	ASTM D445	5.5–7.5 cSt
Kinematic carbon number	ASTM D5442	≥80% C26–C50
Distribution of normal hydrocarbons		<20% C26 and below <10% above C50 Zero above C85
Non-normal paraffin	ASTM D5442	10–50 % Content
Needle penetration (77°F)	ASTM D1321	10–18
Oil content	ASTM D721	≤1%
Flash point	ASTM D92	446°F (230°C)
Color (Saybolt)	ASTM D156	≥10
Acid number	ASTM D1386	≤0.5
Density	ASTM D792	0.915–0.940 g/cm^3
Physical appearance		Small uniform flake prill or powder‡

* Test methods not prescribed shall be those currently used in the industry.
† Hydrocarbon waxes containing linear and branched chains with carbon numbers from C20 to C60.
‡ This requirement is not applicable when the wax is added as a liquid.

Hydrocarbon waxes not meeting the requirements in Table A-7 can qualify as a functionally equivalent ingredient upon successful completion of Sec. A2.3 testing requirements.

Note: The above requirements are not specifications or standards for hydrocarbon waxes. They do not describe all the commercial hydrocarbon waxes that are suitable for use in PVC pipe compounds. These requirements only serve to define members within a group of commercial hydrocarbon waxes that are apparently chemically identical and that have been found by physical/chemical analysis, by pipe testing, and by use of pipe in the field to be sufficiently similar in nature to produce pipe and fittings that exhibit basically the same properties and performance.

Table A-8 Polyethylene wax property requirements

Property	Test Method	Requirement
Gardner color (Molten wax)	ASTM D1544	≤4
Flash point, minimum	ASTM D92	500°F (260° C) open cup 425°F (218° C) closed cup
Ring and ball softening point	ASTM E28	210–225°F (99–107 °C)
or		
Mettler drop point test	ASTM D3954	203–221°F (95–105 °C)
Thermosel viscosity at 283°F (140°C)	ASTM D3236	70–400 cP
Penetration hardness	ASTM D5	3–6
Acid number, mg KOH/g	*	10–18

* The test method may be any commonly used by the industry such as Allied Chemical W-305-TW-1 or Eastman Chemical ECD 27-102.

A2.2.4 Polyethylene wax. Commercial polyethylene waxes may be substituted one for another without additional stress rupture testing provided: (1) the substitution is for the same amount, (2) the substitution in the compound is 0.3 parts or less per hundred parts of resin, and (3) the properties fall within the ranges given in Table A-8 or the ingredient has been qualified as functionally equivalent.

Polyethylene waxes not meeting the requirements in Table A-8 can qualify as a functionally equivalent ingredient upon successful completion of Sec. A2.3 testing requirements.

Note: The above requirements are not specifications or standards for polyethylene waxes. They do not describe all the commercial polyethylene waxes that are suitable for use in PVC pipe compounds. These requirements only serve to define members within a group of commercial polyethylene waxes that are apparently chemically identical and that have been found by physical/chemical analysis, by pipe testing, and by use of pipe in the field to be sufficiently similar in nature to produce pipe and fittings that exhibit basically the same properties and performance.

A2.2.5 Titanium dioxide. Commercial titanium dioxides may be substituted one for another without additional stress rupture testing provided: (1) the substitution is for the same amount, (2) the substitution in the compound is three parts or less per hundred parts of resin, and (3) they meet the requirements given in Table A-9 or the ingredient has been qualified as functionally equivalent.

Titanium dioxides not meeting the requirements in Table A-9 can qualify as a functionally equivalent ingredient upon successful completion of Sec. A2.3 testing requirements.

Note: The above requirements are not specifications or standards for titanium dioxides. They do not describe all the commercial titanium dioxides that are suitable for use in PVC pipe compounds. These requirements only serve to define members within a group of commercial titanium dioxides that are apparently chemically identical and that have been found by physical/chemical analysis, by pipe testing, and by use of pipe in the field to be sufficiently similar in nature to produce pipe and fittings that exhibit basically the same properties and performance.

A2.2.6 Calcium carbonate. The calcium carbonate particles may be either uncoated or coated with calcium stearate and/or stearic acid. Substitutions between uncoated and coated calcium carbonates shall not be permitted without confirmatory stress rupture testing.

A2.2.6.1 Uncoated calcium carbonate. Uncoated commercial calcium carbonates may be substituted one for another without additional stress rupture testing provided: (1) the substitution is for the same amount, (2) the amount used in the compound is five parts

Table A-9 Titanium dioxide property requirements

Property*	Requirement
Crystal structure	Rutile
Particle size, average Retained on 325 mesh screen	0.10–0.35 microns ≤0.20%
Titanium dioxide content	≥92%
Chemical modification	Alumina and/or silica
Carbon content at 1625°F (900°C)†	≤0.30%
Volatile content loss at 221°F (105°C)	≤0.70%
Specific gravity	4.0–4.3

* The test methods shall be those described in ASTM D476, Standard Specification for Titanium Dioxide Pigments.
† Burn and measure carbon dioxide.

Table A-10 Uncoated calcium carbonate property requirements

Property*	Requirement
Calcium carbonate content	≥94.00%
Iron as ferric oxide content	≤0.25%
Silica (SiO_2) content	≤1.25%
Weight loss at 221°F (105°C)	≤0.50%
Particle size range	0.02–15.00 microns
Mean particle size	0.05–3.50 microns
Density	2.65–2.71 g/cm^3

* The required test methods are given in Table A-12.

or less per hundred parts of resin, and (3) the properties fall within the ranges given in Table A-10 or the ingredient has been qualified as functionally equivalent.

A2.2.6.2 Coated calcium carbonate. Coated calcium carbonates shall have particles coated with commercial grades of calcium stearate and/or stearic acid. The fatty acid mixtures in these stearates shall be over 50 percent by weight stearic acid with other fatty acids, mainly palmitic and a small amount of myristic. These stearate-coated commercial calcium carbonates may be substituted one for another without additional stress rupture testing provided: (1) the substitution is for the same amount, (2) the amount used in the compound is five parts or less per hundred parts of resin, and (3) the properties fall within the ranges given in Table A-11 or the ingredient has been qualified as functionally equivalent.

Uncoated or coated calcium carbonates not meeting the requirements in Table A-10 or A-11 can qualify as a functionally equivalent ingredient upon successful completion of Sec. A2.3 testing requirements.

Note: The previously described requirements are not specifications or standards for uncoated and coated calcium carbonates. They do not describe all the commercial calcium carbonates that are suitable for use in PVC pipe compounds. These requirements only serve to define members within a group of commercial uncoated and coated calcium carbonates that are apparently chemically identical and that have been found by physical/chemical analysis, by pipe testing, and by use of pipe in the field to be sufficiently similar in nature to produce pipe and fittings that exhibit basically the same properties and performance.

Table A-11 Stearate-coated calcium carbonate property requirements

Property*	Requirement
Calcium carbonate analysis	≥93.0%
Iron as ferric oxide	≤0.25%
Silica SiO_2	≤1.25%
Weight loss at 221°F (105°C)	≤0.50%
Fatty acids as stearic acid	0.75–2.00%
Particle size range	0.02–15.00 microns
Mean particle size	0.05–3.50 microns
Density	2.65–2.71 g/cm^3

* The required test methods are given in Table A-12.

Table A-12 Required test procedures and references for the evaluation of calcium carbonate properties

Calcium carbonate analysis: Data shall be generated using EDTA titration procedures as per ASTM-C25 or equivalent, or as listed in the assay procedures of the current edition of Food Chemicals Codex for Limestone, Ground.

Iron as ferric oxide: Data shall be generated using wet chemical analysis as per ASTM-C25 or equivalent, or when using firmly established Atomic Absorption methods.

Silica, SiO_2: Data shall be generated using wet chemical analysis as per ASTM-C25 or equivalent or using firmly established X-ray diffraction methodology.

Weight loss at 221°F: Data shall be generated using an oven method as described in ASTM-C25 or equivalent.

Fatty acids as stearic acid: Data shall be generated using a heated weight loss procedure or by using firmly established thermo gravimetric analysis (TGA) methods.

Particle type: Visual inspection via optical or SEM microscopy shall be used.

Particle size, range, and mean: Data shall be generated using a sedigraph and following procedures as described in ASTM-D1199 or equivalent.

Density: Data shall be generated following procedures as described in ASTM-D153 and C188 or their equivalents.

Sec. A2.3. Requirements for Qualification as a Functionally Equivalent Ingredient

Any ingredient not meeting the allowable content range and the physical property requirements given in Section A2.2 can be qualified as a functionally equivalent ingredient when sufficient ASTM D2837 stress-rupture test data confirm the HDB rating of 4,000 psi at 73°F (23°C). Three lots of pipe extruded on commercial equipment from compounds that fall within the ingredient range limits listed in Table A-5, with the subject ingredient at the maximum proposed use level, shall be tested. In cases where the minimum proposed use level of the subject ingredient is greater than zero, one lot shall include the subject ingredient at the maximum proposed use level and another lot shall include the subject ingredient at the minimum proposed use level. Each test lot shall be formulated with a different, qualified PVC resin.

A2.3.1 The ingredient shall be eligible for initial qualification as a functional equivalent upon successful completion of I-2 level testing on one lot. The initial qualification testing, together with testing of two additional lots, must be advanced and completed within one year.

A2.3.2 Qualification as a functionally equivalent ingredient with no expiration date shall require successful completion of I-10 level testing for one lot and I-2 level data for two other lots as required in Sec A2.3,

Sec. A2.4 Compliance Verification

The purchaser may require an affidavit from the manufacturer, supplier, or a third-party that an ingredient or combination of ingredients complies with the applicable requirements of this Appendix.

REFERENCES

ASTM International. 2013. D2837. Standard Test Method for Obtaining Hydrostatic Design Basis for Thermoplastic Pipe Materials or Pressure Design Basis for Thermoplastic Pipe Products. West Conshohocken, Pa: ASTM International.

Plastics Pipe Institute. 2018. PVC Range Composition Listing of Qualified Ingredients. PPI Technical Report, PPI-TR-2. Irving, Tex.: Plastics Pipe Institute.

Plastics Pipe Institute. 2018. Policies and Procedures for Developing Recommended Hydrostatic Design Stresses for Thermoplastic Pipe Materials. PPI Technical Report, PPI-TR-3. Irving, Tex.: Plastic Pipe Institute.

Appendix **B**

Hydraulic Tables

Foreword

This appendix to AWWA Manual 23 includes useful tables for hydraulic design specific to PVC pipe.

Table B-1.A Flow friction loss, AWWA C900 4-in. CIOD PVC pipe

4-in. CIOD									
	DR 25 Pressure class 165 psi, D_i = 4.39 in.			DR 18 Pressure class 235 psi, D_i = 4.23 in.			DR 14 Pressure class 305 psi, D_i = 4.07 in.		
		Pressure drop			Pressure drop			Pressure drop	
Flow, gpm	Velocity, ft/s	ft H_2O/100 ft	psi/100 ft	Velocity, ft/s	ft H_2O/100 ft	psi/100 ft	Velocity, ft/s	ft H_2O/100 ft	psi/100 ft
25	0.530	0.0286	0.0124	0.571	0.0343	0.0148	0.617	0.0414	0.0179
40	0.849	0.0683	0.0296	0.914	0.0818	0.0354	0.987	0.0987	0.0427
55	1.17	0.123	0.0533	1.26	0.147	0.0638	1.36	0.178	0.0770
70	1.49	0.192	0.0833	1.60	0.230	0.0997	1.73	0.278	0.120
85	1.80	0.275	0.119	1.94	0.330	0.143	2.10	0.398	0.172
100	2.12	0.372	0.161	2.29	0.446	0.193	2.47	0.537	0.233
120	2.55	0.521	0.226	2.74	0.624	0.270	2.96	0.753	0.326
140	2.97	0.693	0.300	3.20	0.830	0.360	3.46	1.00	0.434
160	3.39	0.888	0.384	3.66	1.06	0.460	3.95	1.28	0.555
180	3.82	1.10	0.478	4.11	1.32	0.572	4.44	1.59	0.690
200	4.24	1.34	0.581	4.57	1.61	0.696	4.94	1.94	0.839
225	4.77	1.67	0.722	5.14	2.00	0.865	5.55	2.41	1.04
250	5.30	2.03	0.878	5.71	2.43	1.05	6.17	2.93	1.27
275	5.83	2.42	1.05	6.28	2.90	1.25	6.79	3.49	1.51
300	6.36	2.84	1.23	6.86	3.40	1.47	7.40	4.10	1.78
350	7.43	3.78	1.64	8.00	4.52	1.96	8.64	5.46	2.36
400	8.49	4.84	2.09	9.14	5.79	2.51	9.87	6.99	3.02
450	9.55	6.01	2.60	10.3	7.20	3.12	11.1	8.69	3.76

Notes:
1. Table is based on Eq 3-11
2. Friction-loss values are based on average $D_i = D_o - (2 \times 106\%\ t_{min}) = D_o - (2.12 \times t_{min})$, where:
   D_i = pipe inside diameter, in.
   D_o = pipe outside diameter, in.
   t_{min} = minimum wall thickness, in.

Table B-1.B Flow friction loss, AWWA C900 6-in. CIOD PVC pipe

	6-in. CIOD								
	DR 25 Pressure class 165 psi, D_i = 6.31 in.			DR 18 Pressure class 235 psi, D_i = 6.09 in.			DR 14 Pressure class 305 psi, D_i = 5.86 in.		
		Pressure drop			Pressure drop			Pressure drop	
Flow, gpm	Velocity, ft/s	ft H_2O/100 ft	psi/100 ft	Velocity, ft/s	ft H_2O/100 ft	psi/100 ft	Velocity, ft/s	ft H_2O/100 ft	psi/100 ft
50	0.513	0.0177	0.00766	0.551	0.0210	0.00911	0.595	0.0254	0.0110
75	0.770	0.0375	0.0162	0.827	0.0445	0.0193	0.893	0.0537	0.0232
100	1.03	0.0638	0.0276	1.10	0.0758	0.0328	1.19	0.0914	0.0396
125	1.28	0.0964	0.0417	1.38	0.115	0.0496	1.49	0.138	0.0598
150	1.54	0.135	0.0585	1.65	0.161	0.0695	1.79	0.194	0.0838
175	1.80	0.180	0.0778	1.93	0.213	0.0924	2.08	0.257	0.111
200	2.05	0.230	0.0996	2.20	0.273	0.118	2.38	0.330	0.143
250	2.57	0.348	0.150	2.76	0.413	0.179	2.98	0.498	0.216
300	3.08	0.487	0.211	3.31	0.579	0.251	3.57	0.698	0.302
350	3.59	0.648	0.280	3.86	0.770	0.333	4.17	0.928	0.402
400	4.11	0.829	0.359	4.41	0.985	0.427	4.76	1.19	0.514
450	4.62	1.03	0.446	4.96	1.23	0.531	5.36	1.48	0.640
500	5.13	1.25	0.543	5.51	1.49	0.645	5.95	1.80	0.777
600	6.16	1.76	0.760	6.61	2.09	0.903	7.14	2.52	1.09
700	7.19	2.33	1.01	7.72	2.77	1.20	8.33	3.35	1.45
800	8.22	2.99	1.29	8.82	3.55	1.54	9.53	4.28	1.85
900	9.24	3.72	1.61	9.92	4.42	1.91	10.7	5.33	2.31
1,000	10.3	4.52	1.96	11.0	5.37	2.32	11.9	6.47	2.80

Table B-1.C Flow friction loss, AWWA C900 8-in. CIOD PVC pipe

	8-in. CIOD								
	DR 25 Pressure class 165 psi, D_i = 8.28 in.			DR 18 Pressure class 235 psi, D_i = 7.98 in.			DR 14 Pressure class 305 psi, D_i = 7.68 in.		
		Pressure drop			Pressure drop			Pressure drop	
Flow, gpm	Velocity, ft/s	ft H_2O/100 ft	psi/100 ft	Velocity, ft/s	ft H_2O/100 ft	psi/100 ft	Velocity, ft/s	ft H_2O/100 ft	psi/100 ft
100	0.596	0.0170	0.00738	0.642	0.0204	0.00883	0.693	0.0246	0.0106
125	0.745	0.0257	0.0111	0.803	0.0308	0.0133	0.866	0.0371	0.0161
150	0.895	0.0361	0.0156	0.963	0.0432	0.0187	1.04	0.0520	0.0225
200	1.19	0.0614	0.0266	1.28	0.0735	0.0318	1.39	0.0885	0.0383
250	1.49	0.0928	0.0402	1.61	0.111	0.0481	1.73	0.134	0.0579
300	1.79	0.130	0.0563	1.93	0.156	0.0674	2.08	0.187	0.0812
350	2.09	0.173	0.0749	2.25	0.207	0.0896	2.43	0.249	0.108
400	2.39	0.221	0.0959	2.57	0.265	0.115	2.77	0.319	0.138
450	2.68	0.275	0.119	2.89	0.329	0.143	3.12	0.397	0.172
500	2.98	0.335	0.145	3.21	0.400	0.173	3.47	0.482	0.209
600	3.58	0.469	0.203	3.85	0.561	0.243	4.16	0.676	0.293
700	4.17	0.623	0.270	4.49	0.746	0.323	4.85	0.899	0.389
800	4.77	0.798	0.346	5.14	0.955	0.413	5.55	1.15	0.498
1,000	5.96	1.21	0.522	6.42	1.44	0.625	6.93	1.74	0.753
1,200	7.16	1.69	0.732	7.70	2.02	0.875	8.32	2.44	1.05
1,400	8.35	2.25	0.973	8.99	2.69	1.16	9.70	3.24	1.40
1,600	9.54	2.88	1.25	10.3	3.44	1.49	11.1	4.15	1.80
1,800	10.7	3.58	1.55	11.6	4.28	1.85	12.5	5.16	2.23

Notes:
1. Table is based on Eq 3-11
2. Friction-loss values are based on average $D_i = D_o - (2 \times 106\%\ t_{min}) = D_o - (2.12 \times t_{min})$, where:
D_i = pipe inside diameter, in.
D_o = pipe outside diameter, in.
t_{min} = minimum wall thickness, in.

Table B-1.D Flow friction loss, AWWA C900 10-in. CIOD PVC pipe

	10-in. CIOD								
	DR 25 Pressure class 165 psi, D_i = 10.16 in.			DR 18 Pressure class 235 psi, D_i = 9.79 in.			DR 14 Pressure class 305 psi, D_i = 9.42 in.		
		Pressure drop			Pressure drop			Pressure drop	
Flow, gpm	Velocity, ft/s	ft H_2O/100 ft	psi/100 ft	Velocity, ft/s	ft H_2O/100 ft	psi/100 ft	Velocity, ft/s	ft H_2O/100 ft	psi/100 ft
150	0.594	0.0133	0.00578	0.640	0.0160	0.00692	0.691	0.0193	0.00834
200	0.792	0.0227	0.00984	0.853	0.0272	0.0118	0.922	0.0328	0.0142
250	0.990	0.0343	0.0149	1.07	0.0411	0.0178	1.15	0.0496	0.0215
300	1.19	0.0481	0.0208	1.28	0.0576	0.0249	1.38	0.0695	0.0301
350	1.39	0.0640	0.0277	1.49	0.0766	0.0332	1.61	0.0924	0.0400
400	1.58	0.0819	0.0355	1.71	0.0981	0.0425	1.84	0.118	0.0512
450	1.78	0.102	0.0441	1.92	0.122	0.0528	2.07	0.147	0.0637
500	1.98	0.124	0.0536	2.13	0.148	0.0642	2.30	0.179	0.0774
600	2.38	0.173	0.0751	2.56	0.208	0.0899	2.76	0.250	0.108
700	2.77	0.231	0.100	2.99	0.276	0.120	3.23	0.333	0.144
800	3.17	0.295	0.128	3.41	0.354	0.153	3.69	0.426	0.185
1,000	3.96	0.446	0.193	4.27	0.534	0.231	4.61	0.644	0.279
1,200	4.75	0.625	0.271	5.12	0.749	0.324	5.53	0.903	0.391
1,400	5.55	0.831	0.360	5.97	1.00	0.431	6.45	1.20	0.520
1,600	6.34	1.06	0.461	6.83	1.27	0.552	7.37	1.54	0.666
1,800	7.13	1.32	0.573	7.68	1.59	0.686	8.29	1.91	0.828
2,200	8.71	1.92	0.831	9.39	2.30	0.995	10.1	2.77	1.20
2,600	10.3	2.61	1.13	11.1	3.13	1.36	12.0	3.77	1.63

Table B-1.E Flow friction loss, AWWA C900 12-in. CIOD PVC pipe

12-in. CIOD									
	DR 25 Pressure class 165 psi, D_i = 12.08 in.			DR 18 Pressure class 235 psi, D_i = 11.65 in.			DR 14 Pressure class 305 psi, D_i = 11.20 in.		
		Pressure drop			Pressure drop			Pressure drop	
Flow, gpm	Velocity, ft/s	ft H_2O/100 ft	psi/100 ft	Velocity, ft/s	ft H_2O/100 ft	psi/100 ft	Velocity, ft/s	ft H_2O/100 ft	psi/100 ft
200	0.560	0.00980	0.00424	0.602	0.0117	0.00506	0.652	0.0141	0.00613
250	0.700	0.0148	0.00641	0.753	0.0177	0.00764	0.815	0.0214	0.00926
300	0.841	0.0207	0.00898	0.904	0.0247	0.0107	0.978	0.0300	0.0130
350	0.981	0.0276	0.0119	1.05	0.0329	0.0142	1.14	0.0398	0.0173
400	1.12	0.0353	0.0153	1.20	0.0421	0.0182	1.30	0.0510	0.0221
500	1.40	0.0534	0.0231	1.51	0.0636	0.0276	1.63	0.0771	0.0334
600	1.68	0.0748	0.0324	1.81	0.0892	0.0386	1.96	0.108	0.0468
700	1.96	0.0994	0.0431	2.11	0.119	0.0514	2.28	0.144	0.0622
800	2.24	0.127	0.0551	2.41	0.152	0.0657	2.61	0.184	0.0796
1,000	2.80	0.192	0.0833	3.01	0.229	0.0993	3.26	0.278	0.120
1,200	3.36	0.270	0.117	3.61	0.321	0.139	3.91	0.389	0.169
1,400	3.92	0.358	0.155	4.22	0.428	0.185	4.56	0.518	0.224
1,600	4.48	0.459	0.199	4.82	0.547	0.237	5.22	0.663	0.287
2,000	5.60	0.694	0.300	6.02	0.827	0.358	6.52	1.00	0.434
2,400	6.72	0.972	0.421	7.23	1.16	0.502	7.82	1.40	0.608
2,800	7.85	1.29	0.560	8.43	1.54	0.667	9.13	1.87	0.808
3,200	8.97	1.65	0.716	9.64	1.97	0.854	10.4	2.39	1.03
3,600	10.1	2.06	0.891	10.8	2.45	1.06	11.7	2.97	1.29

Notes:

1. Table is based on Eq 3-11
2. Friction-loss values are based on average $D_i = D_o - (2 \times 106\%\ t_{min}) = D_o - (2.12 \times t_{min})$, where:

   D_i = pipe inside diameter, in.

   D_o = pipe outside diameter, in.

   t_{min} = minimum wall thickness, in.

Table B-1.F Flow friction loss, AWWA C900 14-in. CIOD PVC pipe

	14-in. CIOD								
	DR 41 Pressure class 100 psi, D_i = 14.51 in.			DR 32.5 Pressure class 125 psi, D_i = 14.30 in.			DR 25 Pressure class 165 psi, D_i = 14.00 in.		
		Pressure drop			Pressure drop			Pressure drop	
Flow, gpm	Velocity, ft/s	ft H_2O/100 ft	psi/100 ft	Velocity, ft/s	ft H_2O/100 ft	psi/100 ft	Velocity, ft/s	ft H_2O/100 ft	psi/100 ft
300	0.583	0.00851	0.00369	0.600	0.00914	0.00396	0.626	0.0101	0.00439
450	0.874	0.0180	0.00780	0.900	0.0193	0.00838	0.939	0.0214	0.00928
600	1.17	0.0307	0.0133	1.20	0.0329	0.0143	1.25	0.0365	0.0158
800	1.55	0.0522	0.0226	1.60	0.0561	0.0243	1.67	0.0622	0.0269
1,000	1.94	0.0789	0.0342	2.00	0.0847	0.0367	2.09	0.0939	0.0407
1,200	2.33	0.111	0.0479	2.40	0.119	0.0514	2.50	0.132	0.0570
1,400	2.72	0.147	0.0637	2.80	0.158	0.0684	2.92	0.175	0.0758
1,600	3.11	0.188	0.0815	3.20	0.202	0.0875	3.34	0.224	0.0970
1,800	3.50	0.234	0.101	3.60	0.251	0.109	3.75	0.279	0.121
2,000	3.88	0.285	0.123	4.00	0.305	0.132	4.17	0.339	0.147
2,200	4.27	0.339	0.147	4.40	0.364	0.158	4.59	0.404	0.175
2,600	5.05	0.462	0.200	5.20	0.496	0.215	5.42	0.550	0.238
3,000	5.83	0.603	0.261	6.00	0.647	0.280	6.26	0.717	0.310
3,400	6.60	0.759	0.329	6.80	0.815	0.353	7.09	0.904	0.391
3,800	7.38	0.933	0.404	7.60	1.00	0.434	7.93	1.11	0.481
4,200	8.16	1.12	0.486	8.40	1.21	0.522	8.76	1.34	0.579
4,600	8.93	1.33	0.575	9.20	1.43	0.618	9.60	1.58	0.685
5,000	9.71	1.55	0.671	10.0	1.66	0.721	10.4	1.84	0.799

Table B-1.F Flow friction loss, AWWA C900 14-in. CIOD PVC pipe *(continued)*

14-in. CIOD									
	DR 21 Pressure class 200 psi, D_i = 13.76 in.			DR 18 Pressure class 235 psi, , D_i = 13.50 in.			DR 14 Pressure class 305 psi, D_i = 12.98 in.		
		Pressure drop			Pressure drop			Pressure drop	
Flow, gpm	Velocity, ft/s	ft H_2O/100 ft	psi/100 ft	Velocity, ft/s	ft H_2O/100 ft	psi/100 ft	Velocity, ft/s	ft H_2O/100 ft	psi/100 ft
300	0.648	0.0110	0.00477	0.673	0.0121	0.00523	0.728	0.0146	0.00633
450	0.972	0.0233	0.0101	1.01	0.0256	0.0111	1.09	0.0310	0.0134
600	1.30	0.0397	0.0172	1.35	0.0436	0.0189	1.46	0.0527	0.0228
800	1.73	0.0676	0.0293	1.79	0.0742	0.0321	1.94	0.0898	0.0389
1,000	2.16	0.102	0.0442	2.24	0.112	0.0485	2.43	0.136	0.0587
1,200	2.59	0.143	0.0620	2.69	0.157	0.0680	2.91	0.190	0.0823
1,400	3.02	0.190	0.0824	3.14	0.209	0.0904	3.40	0.253	0.109
1,600	3.46	0.244	0.106	3.59	0.267	0.116	3.88	0.324	0.140
1,800	3.89	0.303	0.131	4.04	0.333	0.144	4.37	0.402	0.174
2,000	4.32	0.368	0.159	4.49	0.404	0.175	4.85	0.489	0.212
2,200	4.75	0.439	0.190	4.94	0.482	0.209	5.34	0.583	0.253
2,600	5.61	0.598	0.259	5.83	0.657	0.284	6.31	0.795	0.344
3,000	6.48	0.780	0.338	6.73	0.856	0.370	7.28	1.04	0.448
3,400	7.34	0.983	0.426	7.63	1.08	0.467	8.25	1.31	0.565
3,800	8.21	1.21	0.523	8.52	1.32	0.574	9.22	1.60	0.694
4,200	9.07	1.45	0.629	9.42	1.59	0.690	10.2	1.93	0.836
4,600	9.93	1.72	0.745	10.3	1.89	0.817	11.2	2.28	0.989
5,000	10.8	2.01	0.869	11.2	2.20	0.953	12.1	2.66	1.15

Notes:
1. Table is based on Eq 3-11
2. Friction-loss values are based on average $D_i = D_o - (2 \times 106\%\ t_{min}) = D_o - (2.12 \times t_{min})$, where:
   D_i = pipe inside diameter, in.
   D_o = pipe outside diameter, in.
   t_{min} = minimum wall thickness, in.

Table B-1.G Flow friction loss, AWWA C900 16-in. CIOD PVC pipe

	16-in. CIOD								
	DR 41 Pressure class 100 psi, D_i = 16.50 in.			DR 32.5 Pressure class 125 psi, D_i = 16.26 in.			DR 25 Pressure class 165 psi, D_i = 15.92 in.		
		Pressure drop			Pressure drop			Pressure drop	
Flow, gpm	Velocity, ft/s	ft H_2O/100 ft	psi/100 ft	Velocity, ft/s	ft H_2O/100 ft	psi/100 ft	Velocity, ft/s	ft H_2O/100 ft	psi/100 ft
350	0.526	0.00606	0.00262	0.541	0.00651	0.00282	0.565	0.00721	0.00312
500	0.751	0.0117	0.00508	0.773	0.0126	0.00545	0.807	0.0140	0.00604
650	0.976	0.0191	0.00825	1.01	0.0205	0.00886	1.05	0.0227	0.00982
800	1.20	0.0280	0.0121	1.24	0.0300	0.0130	1.29	0.0333	0.0144
1,000	1.50	0.0423	0.0183	1.55	0.0454	0.0197	1.61	0.0503	0.0218
1,200	1.80	0.0592	0.0256	1.86	0.0636	0.0275	1.94	0.0705	0.0305
1,400	2.10	0.0788	0.0341	2.17	0.0846	0.0366	2.26	0.0937	0.0406
1,800	2.70	0.125	0.0543	2.78	0.135	0.0583	2.90	0.149	0.0646
2,200	3.30	0.182	0.0787	3.40	0.195	0.0845	3.55	0.216	0.0937
2,600	3.90	0.248	0.107	4.02	0.266	0.115	4.19	0.295	0.128
3,000	4.51	0.323	0.140	4.64	0.346	0.150	4.84	0.384	0.166
3,500	5.26	0.429	0.186	5.41	0.461	0.200	5.65	0.511	0.221
4,000	6.01	0.549	0.238	6.19	0.590	0.255	6.45	0.654	0.283
4,500	6.76	0.683	0.296	6.96	0.733	0.318	7.26	0.813	0.352
5,000	7.51	0.830	0.359	7.73	0.891	0.386	8.07	0.988	0.428
5,500	8.26	0.990	0.429	8.51	1.06	0.460	8.87	1.18	0.510
6,000	9.01	1.16	0.504	9.28	1.25	0.541	9.68	1.38	0.599
6,500	9.76	1.35	0.584	10.1	1.45	0.627	10.5	1.60	0.695

Table B-1.G Flow friction loss, AWWA C900 16-in. CIOD PVC pipe *(continued)*

16-in. CIOD									
	DR 21 Pressure class 200 psi, D_i = 15.64 in.			DR 18 Pressure class 235 psi, D_i = 15.35 in.			DR 14 Pressure class 305 psi, D_i = 14.77 in.		
		Pressure drop			Pressure drop			Pressure drop	
Flow, gpm	Velocity, ft/s	ft H_2O/100 ft	psi/100 ft	Velocity, ft/s	ft H_2O/100 ft	psi/100 ft	Velocity, ft/s	ft H_2O/100 ft	psi/100 ft
350	0.585	0.00786	0.00340	0.607	0.00861	0.00373	0.656	0.0104	0.00450
500	0.836	0.0152	0.00659	0.868	0.0167	0.00721	0.937	0.0201	0.00870
650	1.09	0.0247	0.0107	1.13	0.0271	0.0117	1.22	0.0326	0.0141
800	1.34	0.0363	0.0157	1.39	0.0397	0.0172	1.50	0.0479	0.0207
1,000	1.67	0.0548	0.0237	1.74	0.0600	0.0260	1.87	0.0724	0.0314
1,200	2.01	0.0768	0.0333	2.08	0.0841	0.0364	2.25	0.101	0.0439
1,400	2.34	0.102	0.0442	2.43	0.112	0.0485	2.62	0.135	0.0584
1,800	3.01	0.163	0.0704	3.12	0.178	0.0771	3.37	0.215	0.0930
2,200	3.68	0.236	0.102	3.82	0.258	0.112	4.12	0.311	0.135
2,600	4.35	0.321	0.139	4.51	0.352	0.152	4.87	0.424	0.184
3,000	5.01	0.418	0.181	5.21	0.458	0.198	5.62	0.553	0.239
3,500	5.85	0.557	0.241	6.07	0.610	0.264	6.56	0.735	0.318
4,000	6.69	0.713	0.309	6.94	0.780	0.338	7.50	0.941	0.407
4,500	7.52	0.886	0.384	7.81	0.970	0.420	8.43	1.17	0.507
5,000	8.36	1.08	0.466	8.68	1.18	0.511	9.37	1.42	0.616
5,500	9.19	1.28	0.556	9.54	1.41	0.609	10.3	1.70	0.734
6,000	10.0	1.51	0.653	10.4	1.65	0.715	11.2	1.99	0.863
6,500	10.9	1.75	0.757	11.3	1.92	0.830	12.2	2.31	1.00

Notes:
1. Table is based on Eq 3-11
2. Friction-loss values are based on average $D_i = D_o - (2 \times 106\%\ t_{min}) = D_o - (2.12 \times t_{min})$, where:
   D_i = pipe inside diameter, in.
   D_o = pipe outside diameter, in.
   t_{min} = minimum wall thickness, in.

Table B-1.H Flow friction loss, AWWA C900 18-in. CIOD PVC pipe

	18-in. CIOD								
	DR 51 Pressure class 80 psi, D_i = 18.69 in.			DR 41 Pressure class 100 psi, D_i = 18.49 in.			DR 32.5 Pressure class 125 psi, D_i = 17.23 in.		
		Pressure drop			Pressure drop			Pressure drop	
Flow, gpm	Velocity, ft/s	ft H_2O/100 ft	psi/100 ft	Velocity, ft/s	ft H_2O/100 ft	psi/100 ft	Velocity, ft/s	ft H_2O/100 ft	psi/100 ft
500	0.585	0.00640	0.00277	0.598	0.00674	0.00292	0.615	0.00722	0.00313
800	0.936	0.0153	0.00661	0.957	0.0161	0.00696	0.984	0.0172	0.00746
1,100	1.29	0.0275	0.0119	1.32	0.0290	0.0126	1.35	0.0311	0.0134
1,400	1.64	0.0430	0.0186	1.67	0.0453	0.0196	1.72	0.0485	0.0210
1,800	2.11	0.0684	0.0296	2.15	0.0721	0.0312	2.21	0.0772	0.0334
2,200	2.58	0.0992	0.0429	2.63	0.105	0.0453	2.71	0.112	0.0485
2,600	3.04	0.135	0.585	3.11	0.142	0.0616	3.20	0.153	0.0660
3,000	3.51	0.176	0.0762	3.59	0.186	0.0803	3.69	0.199	0.0860
3,500	4.10	0.234	0.101	4.19	0.247	0.107	4.31	0.264	0.114
4,000	4.68	0.300	0.130	4.78	0.316	0.137	4.92	0.338	0.147
4,500	5.27	0.373	0.161	5.38	0.393	0.170	5.54	0.421	0.182
5,000	5.85	0.453	0.196	5.98	0.477	0.207	6.15	0.511	0.221
5,500	6.44	0.540	0.234	6.58	0.569	0.247	6.77	0.610	0.264
6,000	6.44	0.635	0.275	7.18	0.669	0.290	7.38	0.716	0.310
6,500	7.61	0.736	0.319	7.77	0.775	0.336	8.00	0.831	0.360
7,000	8.19	0.844	0.366	8.37	0.889	0.385	8.61	0.953	0.413
7,500	8.78	0.959	0.415	8.97	1.01	0.438	9.23	1.08	0.469
8,000	9.36	1.08	0.468	9.57	1.14	0.493	9.84	1.22	0.528

Table B-1.H Flow friction loss, AWWA C900 18-in. CIOD PVC pipe *(continued)*

18-in. CIOD									
	DR 25 Pressure class 165 psi, D_i = 17.85 in.			DR 21 Pressure class 200 psi, D_i = 17.53 in.			DR 18 Pressure class 235 psi, D_i = 17.20 in.		
		Pressure drop			Pressure drop			Pressure drop	
Flow, gpm	Velocity, ft/s	ft H_2O/100 ft	psi/100 ft	Velocity, ft/s	ft H_2O/100 ft	psi/100 ft	Velocity, ft/s	ft H_2O/100 ft	psi/100 ft
500	0.642	0.00800	0.00346	0.665	0.00874	0.00378	0.691	0.00958	0.00415
800	1.03	0.0191	0.00826	1.06	0.0208	0.00902	1.11	0.0229	0.00990
1,100	1.41	0.0344	0.0149	1.46	0.0376	0.0163	1.52	0.0412	0.0178
1,400	1.80	0.0537	0.0233	1.86	0.0587	0.0254	1.93	0.0644	0.0279
1,800	2.31	0.0856	0.0370	2.39	0.0934	0.0405	2.49	0.102	0.0444
2,200	2.82	0.124	0.0537	2.93	0.135	0.0586	3.04	0.149	0.0643
2,600	3.34	0.169	0.0732	3.46	0.184	0.0799	3.59	0.202	0.0876
3,000	3.85	0.220	0.0953	3.99	0.240	0.104	4.15	0.264	0.114
3,500	4.49	0.293	0.127	4.66	0.320	0.138	4.84	0.351	0.152
4,000	5.13	0.375	0.162	5.32	0.409	0.177	5.53	0.449	0.194
4,500	5.77	0.466	0.202	5.99	0.509	0.220	6.22	0.558	0.242
5,000	6.42	0.566	0.245	6.65	0.618	0.268	6.91	0.678	0.294
5,500	7.06	0.676	0.293	7.32	0.738	0.319	7.60	0.809	0.350
6,000	7.70	0.794	0.344	7.98	0.867	0.375	8.29	0.950	0.412
6,500	8.34	0.920	0.398	8.65	1.00	0.435	8.98	1.10	0.477
7,000	8.98	1.06	0.457	9.31	1.15	0.499	9.67	1.26	0.547
7,500	9.62	1.20	0.519	9.98	1.31	0.567	10.4	1.44	0.622
8,000	10.3	1.35	0.585	10.6	1.48	0.639	11.1	1.62	0.701

Notes:

1. Table is based on Eq 3-11
2. Friction-loss values are based on average $D_i = D_o - (2 \times 106\%\ t_{min}) = D_o - (2.12 \times t_{min})$, where:
   D_i = pipe inside diameter, in.
   D_o = pipe outside diameter, in.
   t_{min} = minimum wall thickness, in.

Table B-1.H Flow friction loss, AWWA C900 18-in. CIOD PVC pipe *(continued)*

18-in. CIOD			
	DR 14 Pressure class 305 psi, D_i = 16.55 in.		
		Pressure drop	
Flow, gpm	Velocity, ft/s	ft H_2O/100 ft	psi/100 ft
500	0.746	0.0116	0.00500
800	1.19	0.0276	0.0119
1,100	1.64	0.0497	0.0215
1,400	2.09	0.0776	0.0336
1,800	2.69	0.124	0.0535
2,200	3.28	0.179	0.0776
2,600	3.88	0.244	0.106
3,000	4.48	0.318	0.138
3,500	5.22	0.423	0.183
4,000	5.97	0.541	0.234
4,500	6.72	0.673	0.291
5,000	7.46	0.818	0.354
5,500	8.21	0.976	0.422
6,000	8.96	1.15	0.496
6,500	9.70	1.33	0.575
7,000	10.4	1.52	0.660
7,500	11.2	1.73	0.750
8,000	11.9	1.95	0.845

Table B-1.I Flow friction loss, AWWA C900 20-in. CIOD PVC pipe

	20-in. CIOD								
	DR 51 Pressure class 80 psi, D_i = 20.70 in.			DR 41 Pressure class 100 psi, D_i = 20.48 in.			DR 32.5 Pressure class 125 psi, D_i = 20.19 in.		
		Pressure drop			Pressure drop			Pressure drop	
Flow, gpm	Velocity, ft/s	ft H_2O/100 ft	psi/100 ft	Velocity, ft/s	ft H_2O/100 ft	psi/100 ft	Velocity, ft/s	ft H_2O/100 ft	psi/100 ft
600	0.573	0.00546	0.00236	0.585	0.00575	0.00249	0.602	0.00616	0.00267
1,000	0.954	0.0140	0.00608	0.975	0.0148	0.00640	1.00	0.0159	0.00686
1,400	1.34	0.0262	0.0113	1.36	0.0276	0.0119	1.40	0.0295	0.0128
1,800	1.72	0.0417	0.0180	1.75	0.0439	0.0190	1.81	0.0470	0.0204
2,200	2.10	0.0604	0.0261	2.14	0.0636	0.0275	2.21	0.0682	0.0295
2,600	2.48	0.0822	0.0356	2.53	0.0866	0.0375	2.61	0.0928	0.0402
3,000	2.86	0.107	0.0464	2.92	0.113	0.0489	3.01	0.121	0.0524
3,500	3.34	0.143	0.0617	3.41	0.150	0.0650	3.51	0.161	0.0697
4,000	3.82	0.182	0.0790	3.90	0.192	0.0832	4.01	0.206	0.0892
4,500	4.29	0.227	0.0982	4.39	0.239	0.103	4.51	0.256	0.111
5,000	4.77	0.276	0.119	4.87	0.290	0.126	5.02	0.311	0.135
5,500	5.25	0.329	0.142	5.36	0.346	0.150	5.52	0.371	0.161
6,000	5.73	0.386	0.167	5.85	0.407	0.176	6.02	0.436	0.189
6,500	6.20	0.448	0.194	6.34	0.472	0.204	6.52	0.506	0.219
7,000	6.68	0.514	0.222	6.82	0.541	0.234	7.02	0.580	0.251
8,000	7.63	0.658	0.285	7.80	0.693	0.300	8.02	0.743	0.322
9,000	8.59	0.818	0.354	8.77	0.862	0.373	9.03	0.923	0.400
10,000	9.54	0.994	0.430	9.75	1.05	0.453	10.0	1.12	0.486

Notes:

1. Table is based on Eq 3-11
2. Friction-loss values are based on average $D_i = D_o - (2 \times 106\%\ t_{min}) = D_o - (2.12 \times t_{min})$, where:

 D_i = pipe inside diameter, in.
 D_o = pipe outside diameter, in.
 t_{min} = minimum wall thickness, in.

Table B-1.I Flow friction loss, AWWA C900 20-in. CIOD PVC pipe *(continued)*

	20-in. CIOD								
	DR 25 Pressure class 165 psi, D_i = 19.77 in.			DR 21 Pressure class 200 psi, D_i = 19.42 in.			DR 18 Pressure class 235 psi, D_i = 19.06 in.		
		Pressure drop			Pressure drop			Pressure drop	
Flow, gpm	Velocity, ft/s	ft H_2O/100 ft	psi/100 ft	Velocity, ft/s	ft H_2O/100 ft	psi/100 ft	Velocity, ft/s	ft H_2O/100 ft	psi/100 ft
600	0.628	0.00682	0.00295	0.650	0.00744	0.00322	0.675	0.00815	0.00353
1,000	1.05	0.0176	0.00760	1.08	0.0191	0.00829	1.13	0.0210	0.00908
1,400	1.46	0.0327	0.0142	1.52	0.0357	0.0155	1.58	0.0391	0.0169
1,800	1.88	0.0521	0.0225	1.95	0.0568	0.0246	2.03	0.0622	0.0269
2,200	2.30	0.0755	0.0327	2.39	0.0823	0.0357	2.48	0.0902	0.0390
2,600	2.72	0.103	0.0445	2.82	0.112	0.0486	2.93	0.123	0.0532
3,000	3.14	0.134	0.0580	3.25	0.146	0.0633	3.38	0.160	0.0693
3,500	3.66	0.178	0.0772	3.79	0.194	0.0842	3.94	0.213	0.0922
4,000	4.18	0.228	0.0988	4.34	0.249	0.108	4.50	0.273	0.118
4,500	4.71	0.284	0.123	4.88	0.309	0.134	5.06	0.339	0.147
5,000	5.23	0.345	0.149	5.42	0.376	0.163	5.63	0.412	0.178
5,500	5.75	0.411	0.178	5.96	0.449	0.194	6.19	0.491	0.213
6,000	6.28	0.483	0.209	6.50	0.527	0.228	6.75	0.577	0.250
6,500	6.80	0.560	0.243	7.05	0.611	0.265	7.32	0.669	0.290
7,000	7.32	0.642	0.278	7.59	0.701	0.303	7.88	0.767	0.332
8,000	8.37	0.822	0.356	8.67	0.897	0.388	9.00	0.982	0.425
9,000	9.41	1.02	0.443	9.76	1.12	0.483	10.1	1.22	0.529
10,000	10.5	1.24	0.538	10.8	1.36	0.587	11.3	1.48	0.643

Table B-1.J Flow friction loss, AWWA C900 24-in. CIOD PVC pipe

24-in. CIOD									
	DR 51 Pressure class 80 psi, D_i = 24.73 in.			DR 41 Pressure class 100 psi, D_i = 24.47 in.			DR 32.5 Pressure class 125 psi, D_i = 24.12 in.		
		Pressure drop			Pressure drop			Pressure drop	
Flow, gpm	Velocity, ft/s	ft H_2O/100 ft	psi/100 ft	Velocity, ft/s	ft H_2O/100 ft	psi/100 ft	Velocity, ft/s	ft H_2O/100 ft	psi/100 ft
800	0.535	0.00391	0.00169	0.546	0.00412	0.00178	0.562	0.00442	0.00191
1,200	0.802	0.00829	0.00359	0.819	0.00872	0.00378	0.843	0.00936	0.00405
1,600	1.07	0.0141	0.00611	1.09	0.0149	0.00643	1.12	0.0159	0.00690
2,000	1.34	0.0213	0.00923	1.37	0.0224	0.00972	1.41	0.0241	0.0104
2,600	1.74	0.0346	0.0150	1.78	0.0365	0.0158	1.83	0.0391	0.0169
3,200	2.14	0.0509	0.0220	2.19	0.0536	0.0232	2.25	0.0574	0.0249
3,800	2.54	0.0699	0.0303	2.59	0.0736	0.0319	2.67	0.0789	0.0342
4,600	3.08	0.100	0.0431	3.14	0.105	0.0454	3.23	0.112	0.0487
5,400	3.61	0.134	0.0580	3.69	0.141	0.0610	3.80	0.151	0.0655
6,200	4.14	0.173	0.0749	4.23	0.182	0.0788	4.36	0.195	0.0845
7,000	4.68	0.216	0.0937	4.78	0.228	0.0987	4.92	0.244	0.106
8,000	5.35	0.277	0.120	5.46	0.292	0.126	5.62	0.313	0.135
9,000	6.02	0.345	0.149	6.15	0.363	0.157	6.33	0.389	0.168
10,000	6.69	0.419	0.181	6.83	0.441	0.191	7.03	0.473	0.205
11,000	7.35	0.499	0.216	7.51	0.526	0.228	7.73	0.564	0.244
12,000	8.02	0.587	0.254	8.19	0.618	0.267	8.43	0.662	0.287
13,500	9.03	0.730	0.316	9.22	0.768	0.333	9.49	0.824	0.357
15,000	10.0	0.887	0.384	10.2	0.933	0.404	10.5	1.00	0.433

Notes:

1. Table is based on Eq 3-11
2. Friction-loss values are based on average $D_i = D_o - (2 \times 106\%\ t_{min}) = D_o - (2.12 \times t_{min})$, where:

D_i = pipe inside diameter, in.
D_o = pipe outside diameter, in.
t_{min} = minimum wall thickness, in.

Table B-1.J Flow friction loss, AWWA C900 24-in. CIOD PVC pipe *(continued)*

24-in. CIOD									
	DR 25 Pressure class 165 psi, D_i = 23.61 in.			DR 21 Pressure class 200 psi, D_i = 23.20 in.			DR 18 Pressure class 235 psi, D_i = 22.76 in.		
		Pressure drop			Pressure drop			Pressure drop	
Flow, gpm	Velocity, ft/s	ft H_2O/100 ft	psi/100 ft	Velocity, ft/s	ft H_2O/100 ft	psi/100 ft	Velocity, ft/s	ft H_2O/100 ft	psi/100 ft
800	0.587	0.00490	0.00212	0.608	0.00534	0.00231	0.631	0.00586	0.00254
1,200	0.880	0.0104	0.00449	0.912	0.0113	0.00489	0.947	0.0124	0.00537
1,600	1.17	0.0177	0.00765	1.22	0.0192	0.00833	1.26	0.0211	0.00915
2,000	1.47	0.0267	0.0116	1.52	0.0291	0.0126	1.58	0.0319	0.0138
2,600	1.91	0.0434	0.0188	1.98	0.0473	0.0205	2.05	0.0519	0.0225
3,200	2.35	0.0637	0.0276	2.43	0.0694	0.0300	2.53	0.0761	0.0330
3,800	2.79	0.0876	0.0379	2.89	0.0954	0.0413	3.00	0.105	0.0453
4,600	3.37	0.125	0.0540	3.49	0.136	0.0588	3.63	0.149	0.0645
5,400	3.96	0.168	0.0726	4.10	0.183	0.0791	4.26	0.200	0.0868
6,200	4.55	0.217	0.0938	4.71	0.236	0.102	4.89	0.259	0.112
7,000	5.13	0.271	0.117	5.32	0.295	0.128	5.52	0.324	0.140
8,000	5.87	0.347	0.150	6.08	0.378	0.164	6.31	0.415	0.180
9,000	6.60	0.432	0.187	6.84	0.470	0.203	7.10	0.516	0.223
10,000	7.33	0.525	0.227	7.60	0.571	0.247	7.89	0.627	0.271
11,000	8.07	0.626	0.271	8.36	0.681	0.295	8.68	0.748	0.324
12,000	8.80	0.735	0.318	9.12	0.800	0.346	9.47	0.878	0.380
13,500	9.90	0.914	0.396	10.3	1.00	0.431	10.7	1.09	0.473
15,000	11.0	1.11	0.481	11.4	1.21	0.524	11.8	1.33	0.575

Table B-1.K Flow friction loss, AWWA C900 30-in. CIOD PVC pipe

30-in. CIOD									
	DR 51 Pressure class 80 psi, D_i = 30.67 in.			DR 41 Pressure class 100 psi, D_i = 30.35 in.			DR 32.5 Pressure class 125 psi, D_i = 29.91 in.		
		Pressure drop			Pressure drop			Pressure drop	
Flow, gpm	Velocity, ft/s	ft H_2O/100 ft	psi/100 ft	Velocity, ft/s	ft H_2O/100 ft	psi/100 ft	Velocity, ft/s	ft H_2O/100 ft	psi/100 ft
1,200	0.522	0.00291	0.00126	0.533	0.00306	0.00133	0.548	0.00329	0.00142
1,800	0.782	0.00616	0.00267	0.799	0.00649	0.00281	0.823	0.00696	0.00301
2,400	1.04	0.0105	0.00454	1.07	0.0110	0.00478	1.10	0.0119	0.00513
3,200	1.39	0.0179	0.00774	1.42	0.0188	0.00814	1.46	0.0202	0.00874
4,000	1.74	0.0270	0.0117	1.78	0.0284	0.0123	1.83	0.0305	0.0132
5,000	2.17	0.0408	0.0177	2.22	0.0429	0.0186	2.29	0.0461	0.0200
6,000	2.61	0.0572	0.0248	2.66	0.0602	0.0260	2.74	0.0646	0.0280
7,000	3.04	0.0760	0.0329	3.11	0.0800	0.0346	3.20	0.0859	0.0372
8,000	3.48	0.0973	0.0421	3.55	0.102	0.0444	3.66	0.110	0.0476
9,000	3.91	0.121	0.0524	3.99	0.127	0.0551	4.11	0.137	0.0592
10,000	4.35	0.147	0.0637	4.44	0.155	0.0670	4.57	0.166	0.0719
12,000	5.22	0.206	0.0892	5.33	0.217	0.0939	5.48	0.233	0.101
14,000	6.09	0.274	0.119	6.21	0.288	0.125	6.40	0.310	0.134
16,000	6.95	0.351	0.152	7.10	0.369	0.160	7.31	0.396	0.172
18,000	7.82	0.436	0.189	7.99	0.459	0.199	8.23	0.493	0.213
20,000	8.69	0.530	0.230	8.88	0.558	0.242	9.14	0.599	0.259
22,000	9.56	0.632	0.274	9.77	0.666	0.288	10.1	0.715	0.309
24,000	10.4	0.743	0.322	10.7	0.782	0.339	11.0	0.839	0.363

Notes:

1. Table is based on Eq 3-11
2. Friction-loss values are based on average $D_i = D_o - (2 \times 106\%\ t_{min}) = D_o - (2.12 \times t_{min})$, where:

   D_i = pipe inside diameter, in.
   D_o = pipe outside diameter, in.
   t_{min} = minimum wall thickness, in.

Table B-1.K Flow friction loss, AWWA C900 30-in. CIOD PVC pipe *(continued)*

	30-in. CIOD								
	DR 25 Pressure class 165 psi, D_i = 29.29 in.			DR 21 Pressure class 200 psi, D_i = 28.77 in.			DR 18 Pressure class 235 psi, D_i = 28.23 in.		
		Pressure drop			Pressure drop			Pressure drop	
Flow, gpm	Velocity, ft/s	ft H_2O/100 ft	psi/100 ft	Velocity, ft/s	ft H_2O/100 ft	psi/100 ft	Velocity, ft/s	ft H_2O/100 ft	psi/100 ft
1,200	0.572	0.00364	0.00158	0.593	0.00397	0.00172	0.616	0.00436	0.00189
1,800	0.858	0.00771	0.00334	0.889	0.00841	0.00364	0.923	0.00922	0.00399
2,400	1.14	0.0131	0.00568	1.19	0.0143	0.00620	1.23	0.0157	0.00680
3,200	1.53	0.0223	0.00968	1.58	0.0244	0.0106	1.64	0.0267	0.0116
4,000	1.91	0.0338	0.0146	1.98	0.0368	0.0160	2.05	0.0404	0.0175
5,000	2.38	0.0510	0.0221	2.47	0.0557	0.0241	2.57	0.0610	0.0264
6,000	2.86	0.0715	0.0310	2.96	0.0780	0.0338	3.08	0.0855	0.0370
7,000	3.34	0.0951	0.0412	3.46	0.104	0.0449	3.59	0.114	0.0493
8,000	3.81	0.122	0.0527	3.95	0.133	0.0575	4.10	0.146	0.0631
9,000	4.29	0.151	0.0656	4.45	0.165	0.0715	4.62	0.181	0.0784
10,000	4.77	0.184	0.0797	4.94	0.201	0.0869	5.13	0.220	0.0953
12,000	5.72	0.258	0.112	5.93	0.281	0.122	6.16	0.308	0.134
14,000	6.67	0.343	0.148	6.92	0.374	0.162	7.18	0.410	0.178
16,000	7.63	0.439	0.190	7.90	0.479	0.207	8.21	0.525	0.227
18,000	8.58	0.546	0.236	8.89	0.595	0.258	9.23	0.653	0.283
20,000	9.53	0.663	0.287	9.88	0.724	0.313	10.3	0.793	0.344
22,000	10.5	0.791	0.343	10.9	0.863	0.374	11.3	0.946	0.410
24,000	11.4	0.929	0.402	11.9	1.01	0.439	12.3	1.11	0.481

Table B-1.L Flow friction loss, AWWA C900 36-in. CIOD PVC pipe

	36-in. CIOD								
	DR 51 Pressure class 80 psi, D_i = 36.71 in.			DR 41 Pressure class 100 psi, D_i = 36.32 in.			DR 32.5 Pressure class 125 psi, D_i = 35.80 in.		
		Pressure drop			Pressure drop			Pressure drop	
Flow, gpm	Velocity, ft/s	ft H_2O/100 ft	psi/100 ft	Velocity, ft/s	ft H_2O/100 ft	psi/100 ft	Velocity, ft/s	ft H_2O/100 ft	psi/100 ft
2,000	0.607	0.00313	0.00135	0.620	0.00329	0.00143	0.638	0.00353	0.00153
3,500	1.06	0.00880	0.00381	1.08	0.00927	0.00402	1.12	0.00995	0.00431
5,000	1.52	0.0170	0.00737	1.55	0.0179	0.00777	1.60	0.0192	0.00833
7,000	2.12	0.0317	0.0137	2.17	0.0334	0.0145	2.23	0.0359	0.0155
9,000	2.73	0.0505	0.0219	2.79	0.0532	0.0230	2.87	0.0571	0.0247
11,000	3.34	0.0732	0.0317	3.41	0.0771	0.0334	3.51	0.0827	0.0358
13,000	3.94	0.0998	0.0432	4.03	0.105	0.0455	4.15	0.113	0.0488
15,000	4.55	0.130	0.0563	4.65	0.137	0.0593	4.79	0.147	0.0636
17,000	5.16	0.164	0.0710	5.27	0.173	0.0747	5.42	0.185	0.0802
19,000	5.76	0.201	0.0872	5.89	0.212	0.0918	6.06	0.227	0.0985
21,000	6.37	0.242	0.105	6.51	0.255	0.110	6.70	0.274	0.118
23,000	6.98	0.287	0.124	7.13	0.302	0.131	7.34	0.324	0.140
25,000	7.58	0.334	0.145	7.75	0.352	0.153	7.98	0.378	0.164
27,000	8.19	0.386	0.167	8.37	0.406	0.176	8.61	0.436	0.189
29,000	8.80	0.440	0.191	8.99	0.464	0.201	9.25	0.497	0.215
31,000	9.41	0.498	0.216	9.61	0.524	0.227	9.89	0.563	0.244
33,000	10.0	0.559	0.242	10.2	0.589	0.255	10.5	0.632	0.273
35,000	10.6	0.623	0.270	10.8	0.656	0.284	11.2	0.704	0.305

Notes:
1. Table is based on Eq 3-11
2. Friction-loss values are based on average $D_i = D_o - (2 \times 106\%\ t_{min}) = D_o - (2.12 \times t_{min})$, where:

D_i = pipe inside diameter, in.
D_o = pipe outside diameter, in.
t_{min} = minimum wall thickness, in.

Table B-1.L Flow friction loss, AWWA C900 36-in. CIOD PVC pipe *(continued)*

	36-in. CIOD								
	DR 25 Pressure class 165 psi, D_i = 35.05 in.			DR 21 Pressure class 200 psi, D_i = 34.43 in.			DR 18 Pressure class 235 psi, D_i = 33.78 in.		
		Pressure drop			Pressure drop			Pressure drop	
Flow, gpm	Velocity, ft/s	ft H_2O/100 ft	psi/100 ft	Velocity, ft/s	ft H_2O/100 ft	psi/100 ft	Velocity, ft/s	ft H_2O/100 ft	psi/100 ft
2,000	0.666	0.00391	0.00170	0.690	0.00427	0.00185	0.716	0.00468	0.00203
3,500	1.16	0.0110	0.00477	1.21	0.0120	0.00521	1.25	0.0132	0.00570
5,000	1.66	0.0213	0.00923	1.72	0.0233	0.0101	1.79	0.0255	0.011
7,000	2.33	0.0397	0.0172	2.41	0.0433	0.0188	2.51	0.0475	0.021
9,000	3.00	0.0633	0.0274	3.10	0.0690	0.0299	3.22	0.0756	0.033
11,000	3.66	0.0917	0.0397	3.79	0.100	0.0433	3.94	0.110	0.047
13,000	4.33	0.125	0.0541	4.48	0.136	0.0590	4.66	0.149	0.065
15,000	4.99	0.163	0.0705	5.17	0.178	0.0769	5.37	0.194	0.084
17,000	5.66	0.205	0.0888	5.86	0.224	0.0969	6.09	0.245	0.106
19,000	6.32	0.252	0.109	6.55	0.275	0.119	6.80	0.301	0.130
21,000	6.99	0.303	0.131	7.24	0.331	0.143	7.52	0.362	0.157
23,000	7.65	0.359	0.155	7.93	0.391	0.169	8.24	0.429	0.186
25,000	8.32	0.419	0.181	8.62	0.457	0.198	8.95	0.500	0.217
27,000	8.99	0.483	0.209	9.31	0.527	0.228	9.67	0.577	0.250
29,000	9.65	0.551	0.239	10.0	0.601	0.260	10.4	0.658	0.285
31,000	10.3	0.623	0.270	10.7	0.680	0.294	11.1	0.745	0.323
33,000	11.0	0.700	0.303	11.4	0.763	0.331	11.8	0.836	0.362
35,000	11.6	0.780	0.338	12.1	0.851	0.369	12.5	0.932	0.404

Table B-1.M Flow friction loss, AWWA C900 42-in. CIOD PVC pipe

	42-in. CIOD								
	DR 51 Pressure class 80 psi, D_i = 42.65 in.			DR 41 Pressure class 100 psi, D_i = 42.20 in.			DR 32.5 Pressure class 125 psi, D_i = 41.60 in.		
		Pressure drop			Pressure drop			Pressure drop	
Flow, gpm	Velocity, ft/s	ft H_2O/100 ft	psi/100 ft	Velocity, ft/s	ft H_2O/100 ft	psi/100 ft	Velocity, ft/s	ft H_2O/100 ft	psi/100 ft
2,500	0.562	0.00228	0.000987	0.574	0.00240	0.00104	0.591	0.00257	0.00111
4,000	0.899	0.00544	0.00235	0.918	0.00573	0.00248	0.945	0.00614	0.00266
5,500	1.24	0.00980	0.00424	1.26	0.0103	0.00447	1.30	0.0111	0.00479
7,000	1.57	0.0153	0.00663	1.61	0.0161	0.00698	1.65	0.0173	0.00748
9,500	2.14	0.0269	0.0117	2.18	0.0284	0.0123	2.24	0.0304	0.0132
12,000	2.70	0.0415	0.0180	2.76	0.0437	0.0189	2.84	0.0468	0.0203
14,500	3.26	0.0589	0.0255	3.33	0.0620	0.0269	3.43	0.0665	0.0288
17,000	3.82	0.0791	0.0342	3.90	0.0832	0.0360	4.02	0.0892	0.0386
20,000	4.50	0.107	0.0462	4.59	0.112	0.0487	4.73	0.121	0.0522
23,000	5.17	0.138	0.0599	5.28	0.146	0.0630	5.43	0.156	0.0676
26,000	5.84	0.173	0.0751	5.97	0.183	0.0791	6.14	0.196	0.0848
29,000	6.52	0.212	0.0919	6.66	0.224	0.0968	6.85	0.240	0.104
32,000	7.19	0.255	0.110	7.35	0.268	0.116	7.56	0.288	0.125
35,000	7.87	0.301	0.130	8.04	0.317	0.137	8.27	0.339	0.147
38,000	8.54	0.350	0.152	8.72	0.369	0.160	8.98	0.395	0.171
42,000	9.44	0.421	0.182	9.64	0.444	0.192	9.92	0.476	0.206
46,000	10.3	0.499	0.216	10.6	0.525	0.227	10.9	0.563	0.244
50,000	11.2	0.582	0.252	11.5	0.612	0.265	11.8	0.657	0.284

Notes:
1. Table is based on Eq 3-11
2. Friction-loss values are based on average $D_i = D_o - (2 \times 106\%\ t_{min}) = D_o - (2.12 \times t_{min})$, where:
   D_i = pipe inside diameter, in.
   D_o = pipe outside diameter, in.
   t_{min} = minimum wall thickness, in.

Table B-1.M Flow friction loss, AWWA C900 42-in. CIOD PVC pipe *(continued)*

42-in. CIOD			
	DR 25 Pressure class 165 psi, D_i = 40.73 in.		
		Pressure drop	
Flow, gpm	Velocity, ft/s	ft H_2O/100 ft	psi/100 ft
2,500	0.616	0.00285	0.00123
4,000	0.986	0.00680	0.00295
5,500	1.36	0.0123	0.00531
7,000	1.73	0.0192	0.00829
9,500	2.34	0.0337	0.0146
12,000	2.96	0.0519	0.0225
14,500	3.57	0.0737	0.0319
17,000	4.19	0.0989	0.0428
20,000	4.93	0.134	0.0578
23,000	5.67	0.173	0.0749
26,000	6.41	0.217	0.0940
29,000	7.15	0.266	0.115
32,000	7.89	0.319	0.138
35,000	8.63	0.376	0.163
38,000	9.37	0.438	0.190
42,000	10.4	0.527	0.228
46,000	11.3	0.624	0.270
50,000	12.3	0.728	0.315

Table B-1.N Flow friction loss, AWWA C900 48-in. CIOD PVC pipe

	48-in. CIOD								
	DR 51 Pressure class 80 psi, D_i = 48.69 in.			DR 41 Pressure class 100 psi, D_i = 48.17 in.			DR 32.5 Pressure class 125 psi, D_i = 47.49 in.		
		Pressure drop			Pressure drop			Pressure drop	
Flow, gpm	Velocity, ft/s	ft H_2O/100 ft	psi/100 ft	Velocity, ft/s	ft H_2O/100 ft	psi/100 ft	Velocity, ft/s	ft H_2O/100 ft	psi/100 ft
3,000	0.517	0.00168	0.000726	0.529	0.00177	0.000765	0.544	0.00189	0.000820
5,000	0.862	0.00432	0.00187	0.881	0.00455	0.00197	0.906	0.00487	0.00211
8,000	1.38	0.0103	0.00446	1.41	0.0108	0.00470	1.45	0.0116	0.00503
11,000	1.90	0.0186	0.00804	1.94	0.0196	0.00847	1.99	0.0210	0.00907
14,000	2.41	0.0290	0.0126	2.47	0.0306	0.0132	2.54	0.0327	0.0142
17,000	2.93	0.0415	0.0180	3.00	0.0438	0.0189	3.08	0.0469	0.0203
20,000	3.45	0.0561	0.0243	3.52	0.0591	0.0256	3.63	0.0633	0.0274
23,000	3.97	0.0726	0.0315	4.05	0.0765	0.0331	4.17	0.0820	0.0355
26,000	4.48	0.0911	0.0395	4.58	0.0960	0.0416	4.71	0.103	0.0446
29,000	5.00	0.112	0.0483	5.11	0.118	0.0509	5.26	0.126	0.0545
32,000	5.52	0.134	0.0579	5.64	0.141	0.0611	5.80	0.151	0.0654
36,000	6.21	0.166	0.0721	6.34	0.175	0.0759	6.53	0.188	0.0813
40,000	6.90	0.202	0.0876	7.05	0.213	0.0923	7.25	0.228	0.0989
44,000	7.59	0.241	0.104	7.75	0.254	0.110	7.98	0.272	0.118
48,000	8.28	0.283	0.123	8.46	0.299	0.129	8.70	0.320	0.139
52,000	8.97	0.329	0.142	9.16	0.346	0.150	9.43	0.371	0.161
56,000	9.66	0.377	0.163	9.87	0.397	0.172	10.2	0.425	0.184
60,000	10.3	0.428	0.185	10.6	0.451	0.195	10.9	0.483	0.209

Notes:
1. Table is based on Eq 3-11
2. Friction-loss values are based on average $D_i = D_o - (2 \times 106\%\ t_{min}) = D_o - (2.12 \times t_{min})$, where:
   D_i = pipe inside diameter, in.
   D_o = pipe outside diameter, in.
   t_{min} = minimum wall thickness, in.

Table B-1.N Flow friction loss, AWWA C900 48-in. CIOD PVC pipe *(continued)*

48-in. CIOD			
	DR 25 Pressure class 165 psi, D_i = 46.49 in.		
		Pressure drop	
Flow, gpm	Velocity, ft/s	ft H_2O/100 ft	psi/100 ft
3,000	0.568	0.00210	0.000909
5,000	0.946	0.00540	0.00234
8,000	1.51	0.0129	0.00558
11,000	2.08	0.0232	0.0101
14,000	2.65	0.0363	0.0157
17,000	3.22	0.0520	0.0225
20,000	3.78	0.0702	0.0304
23,000	4.35	0.0910	0.0394
26,000	4.92	0.114	0.0494
29,000	5.49	0.140	0.0605
32,000	6.05	0.168	0.0726
36,000	6.81	0.208	0.0902
40,000	7.57	0.253	0.110
44,000	8.32	0.302	0.131
48,000	9.08	0.355	0.154
52,000	9.84	0.411	0.178
56,000	10.6	0.472	0.204
60,000	11.4	0.536	0.232

Table B-2.A Flow friction loss, AWWA C900 (and ASTM D2241) 14-in. IPS OD PVC pipe

14-in. IPS									
	DR 41 Pressure rated 100 psi, D_i = 13.28 in.			DR 32.5 Pressure rated 125 psi, D_i = 13.09 in.			DR 26 Pressure rated 160 psi, D_i = 12.86 in.		
		Pressure drop			Pressure drop			Pressure drop	
Flow, gpm	Velocity, ft/s	ft H_2O/100 ft	psi/100 ft	Velocity, ft/s	ft H_2O/100 ft	psi/100 ft	Velocity, ft/s	ft H_2O/100 ft	psi/100 ft
250	0.580	0.00936	0.00405	0.597	0.0100	0.00434	0.618	0.0109	0.00473
300	0.696	0.0131	0.00568	0.716	0.0140	0.00608	0.742	0.0153	0.00663
350	0.812	0.0174	0.00755	0.835	0.0187	0.00809	0.866	0.0204	0.00882
400	0.928	0.0223	0.00966	0.955	0.0239	0.0104	0.989	0.0261	0.0113
500	1.16	0.0337	0.0146	1.19	0.0361	0.0156	1.24	0.0394	0.0171
600	1.39	0.0473	0.0205	1.43	0.0506	0.0219	1.48	0.0552	0.0239
700	1.62	0.0628	0.0272	1.67	0.0674	0.0292	1.73	0.0734	0.0318
800	1.86	0.0805	0.0348	1.91	0.0862	0.0373	1.98	0.0940	0.0407
1,000	2.32	0.122	0.0526	2.39	0.130	0.0564	2.47	0.142	0.0615
1,200	2.78	0.170	0.0738	2.86	0.183	0.0790	2.97	0.199	0.0862
1,400	3.25	0.227	0.0981	3.34	0.243	0.105	3.46	0.265	0.115
1,600	3.71	0.290	0.126	3.82	0.311	0.135	3.96	0.339	0.147
2,000	4.64	0.438	0.190	4.77	0.470	0.203	4.95	0.512	0.222
2,400	5.57	0.614	0.266	5.73	0.658	0.285	5.93	0.717	0.311
2,800	6.50	0.817	0.354	6.68	0.875	0.379	6.92	0.954	0.413
3,200	7.42	1.05	0.453	7.64	1.12	0.485	7.91	1.22	0.529
3,700	8.58	1.37	0.592	8.83	1.47	0.635	9.15	1.60	0.692
4,200	9.74	1.73	0.749	10.0	1.85	0.802	10.4	2.02	0.875

Notes:
1. Table is based on Eq 3-11
2. Friction-loss values are based on average $D_i = D_o - (2 \times 106\%\ t_{min}) = D_o - (2.12 \times t_{min})$, where:
   D_i = pipe inside diameter, in.
   D_o = pipe outside diameter, in.
   t_{min} = minimum wall thickness, in.

Table B-2.A Flow friction loss, AWWA C900 (and ASTM D2241) 14-in. IPS OD PVC pipe *(continued)*

14-in. IPS						
	DR 21 Pressure rated 200 psi, D_i = 12.59 in.			DR 17 Pressure rated 250 psi, D_i = 12.25 in.		
		Pressure drop			Pressure drop	
Flow, gpm	Velocity, ft/s	ft H_2O/100 ft	psi/100 ft	Velocity, ft/s	ft H_2O/100 ft	psi/100 ft
250	0.645	0.0121	0.00525	0.681	0.0138	0.00598
300	0.774	0.0170	0.00735	0.817	0.0193	0.00837
350	0.903	0.0226	0.00978	0.953	0.0257	0.0111
400	1.03	0.0289	0.0125	1.09	0.0329	0.0143
500	1.29	0.0437	0.0189	1.36	0.0498	0.0215
600	1.55	0.0612	0.0265	1.63	0.0697	0.0302
700	1.81	0.0814	0.0352	1.91	0.0927	0.0401
800	2.06	0.104	0.0451	2.18	0.119	0.0514
1,000	2.58	0.157	0.0682	2.72	0.179	0.0777
1,200	3.10	0.221	0.0955	3.27	0.251	0.109
1,400	3.61	0.293	0.127	3.81	0.334	0.145
1,600	4.13	0.376	0.163	4.36	0.428	0.185
2,000	5.16	0.568	0.246	5.44	0.647	0.280
2,400	6.19	0.795	0.344	6.53	0.906	0.392
2,800	7.22	1.06	0.458	7.62	1.21	0.522
3,200	8.26	1.35	0.586	8.71	1.54	0.668
3,700	9.55	1.77	0.767	10.1	2.02	0.874
4,200	10.8	2.24	0.970	11.4	2.55	1.10

Table B-2.B Flow friction loss, AWWA C900 (and ASTM D2241) 16-in. IPS OD PVC pipe

16-in. IPS									
	DR 41 Pressure rated 100 psi, D_i = 15.33 in.			DR 32.5 Pressure rated 125 psi, D_i = 14.96 in.			DR 26 Pressure rated 160 psi, D_i = 14.70 in.		
		Pressure drop			Pressure drop			Pressure drop	
Flow, gpm	Velocity, ft/s	ft H_2O/100 ft	psi/100 ft	Velocity, ft/s	ft H_2O/100 ft	psi/100 ft	Velocity, ft/s	ft H_2O/100 ft	psi/100 ft
300	0.533	0.00685	0.00297	0.548	0.00735	0.00318	0.568	0.00800	0.00346
400	0.710	0.0117	0.00505	0.731	0.0125	0.00542	0.757	0.0136	0.00590
500	0.888	0.0176	0.00763	0.914	0.0189	0.00818	0.947	0.0206	0.00891
600	1.07	0.0247	0.0107	1.10	0.0265	0.0115	1.14	0.0288	0.0125
700	1.24	0.0328	0.0142	1.28	0.0352	0.0152	1.33	0.0384	0.0166
800	1.42	0.0420	0.0182	1.46	0.0451	0.0195	1.51	0.0491	0.0213
1,000	1.78	0.0635	0.0275	1.83	0.0681	0.0295	1.89	0.0742	0.0321
1,300	2.31	0.103	0.0447	2.38	0.111	0.0479	2.46	0.121	0.0522
1,600	2.84	0.152	0.0656	2.92	0.163	0.0704	3.03	0.177	0.0767
1,900	3.37	0.208	0.0902	3.47	0.223	0.0967	3.60	0.243	0.105
2,200	3.91	0.273	0.118	4.02	0.293	0.127	4.17	0.319	0.138
2,500	4.44	0.346	0.150	4.57	0.371	0.161	4.73	0.404	0.175
3,000	5.33	0.485	0.210	5.48	0.520	0.225	5.68	0.566	0.245
3,500	6.22	0.645	0.279	6.40	0.692	0.299	6.63	0.753	0.326
4,000	7.10	0.826	0.358	7.31	0.885	0.383	7.57	0.965	0.418
4,500	7.99	1.03	0.445	8.23	1.10	0.477	8.52	1.20	0.519
5,000	8.88	1.25	0.540	9.14	1.34	0.579	9.47	1.46	0.631
5,500	9.77	1.49	0.644	10.1	1.60	0.691	10.4	1.74	0.753

Notes:

1. Table is based on Eq 3-11
2. Friction-loss values are based on average $D_i = D_o - (2 \times 106\%\ t_{min}) = D_o - (2.12 \times t_{min})$, where:

D_i = pipe inside diameter, in.
D_o = pipe outside diameter, in.
t_{min} = minimum wall thickness, in.

Table B-2.B Flow friction loss, AWWA C900 (and ASTM D2241) 16-in. IPS OD PVC pipe *(continued)*

16-in. IPS						
	DR 21 Pressure rated 200 psi, D_i = 14.38 in.			DR 17 Pressure rated 250 psi, D_i = 14.00 in.		
		Pressure drop			Pressure drop	
Flow, gpm	Velocity, ft/s	ft H_2O/100 ft	psi/100 ft	Velocity, ft/s	ft H_2O/100 ft	psi/100 ft
300	0.593	0.00888	0.00384	0.625	0.0101	0.00438
400	0.790	0.0151	0.00654	0.834	0.0172	0.00745
500	0.988	0.0228	0.00989	1.04	0.0260	0.0113
600	1.19	0.0320	0.0139	1.25	0.0364	0.0158
700	1.38	0.0426	0.0184	1.46	0.0485	0.0210
800	1.58	0.0545	0.0236	1.67	0.0621	0.0269
1,000	1.98	0.0823	0.0357	2.08	0.0938	0.0406
1,300	2.57	0.134	0.0579	2.71	0.152	0.0660
1,600	3.16	0.196	0.0851	3.34	0.224	0.0969
1,900	3.75	0.270	0.117	3.96	0.307	0.133
2,200	4.35	0.354	0.153	4.59	0.403	0.175
2,500	4.94	0.449	0.194	5.21	0.511	0.221
3,000	5.93	0.628	0.272	6.25	0.716	0.310
3,500	6.92	0.836	0.362	7.30	0.952	0.412
4,000	7.90	1.07	0.463	8.34	1.22	0.528
4,500	8.89	1.33	0.576	9.38	1.52	0.656
5,000	9.88	1.62	0.700	10.4	1.84	0.797
5,500	10.9	1.93	0.835	11.5	2.20	0.951

Table B-2.C Flow friction loss, AWWA C900 (and ASTM D2241) 18-in. IPS OD PVC pipe

18-in. IPS									
	DR 41 Pressure rated 100 psi, D_i = 17.25 in.			DR 32.5 Pressure rated 125 psi, D_i = 16.83 in.			DR 26 Pressure rated 160 psi, D_i = 16.53 in.		
		Pressure drop			Pressure drop			Pressure drop	
Flow, gpm	Velocity, ft/s	ft H_2O/100 ft	psi/100 ft	Velocity, ft/s	ft H_2O/100 ft	psi/100 ft	Velocity, ft/s	ft H_2O/100 ft	psi/100 ft
400	0.561	0.00658	0.00285	0.578	0.00706	0.00306	0.598	0.00769	0.00333
600	0.842	0.0139	0.00603	0.867	0.0149	0.00647	0.898	0.0163	0.00705
800	1.12	0.0237	0.0103	1.16	0.0254	0.0110	1.20	0.0277	0.0120
1,000	1.40	0.0358	0.0155	1.44	0.0384	0.0166	1.50	0.0419	0.0181
1,200	1.68	0.0502	0.0217	1.73	0.0539	0.0233	1.80	0.0587	0.0254
1,400	1.96	0.0668	0.0289	2.02	0.0716	0.0310	2.09	0.0780	0.0338
1,800	2.53	0.106	0.0460	2.60	0.114	0.0494	2.69	0.124	0.0538
2,200	3.09	0.154	0.0667	3.18	0.165	0.0716	3.29	0.180	0.0780
2,600	3.65	0.210	0.0909	3.75	0.225	0.0975	3.89	0.245	0.106
3,000	4.21	0.274	0.118	4.33	0.293	0.127	4.49	0.320	0.138
3,500	4.91	0.364	0.158	5.05	0.390	0.169	5.24	0.425	0.184
4,000	5.61	0.466	0.202	5.78	0.500	0.216	5.98	0.544	0.236
4,500	6.31	0.579	0.251	6.50	0.621	0.269	6.73	0.677	0.293
5,000	7.02	0.704	0.305	7.22	0.755	0.327	7.48	0.822	0.356
5,500	7.72	0.840	0.364	7.94	0.900	0.390	8.23	0.981	0.425
6,000	8.42	0.986	0.427	8.67	1.06	0.458	8.98	1.15	0.499
6,500	9.12	1.14	0.495	9.39	1.23	0.531	9.72	1.34	0.578
7,000	9.82	1.31	0.568	10.1	1.41	0.609	10.5	1.53	0.663

Notes:

1. Table is based on Eq 3-11
2. Friction-loss values are based on average $D_i = D_o - (2 \times 106\%\ t_{min}) = D_o - (2.12 \times t_{min})$, where:

D_i = pipe inside diameter, in.
D_o = pipe outside diameter, in.
t_{min} = minimum wall thickness, in.

Table B-2.C Flow friction loss, AWWA C900 (and ASTM D2241) 18-in. IPS OD PVC pipe *(continued)*

18-in. IPS						
	DR 21 Pressure rated 200 psi, D_i = 16.18 in.			DR 17 Pressure rated 250 psi, D_i = 15.76 in.		
		Pressure drop			Pressure drop	
Flow, gpm	Velocity, ft/s	ft H_2O/100 ft	psi/100 ft	Velocity, ft/s	ft H_2O/100 ft	psi/100 ft
400	0.624	0.00853	0.00369	0.659	0.00971	0.00421
600	0.937	0.0181	0.00782	0.988	0.0206	0.00890
800	1.25	0.0307	0.0133	1.32	0.0350	0.0152
1,000	1.56	0.0465	0.0201	1.65	0.0529	0.0229
1,200	1.87	0.0651	0.0282	1.98	0.0741	0.0321
1,400	2.19	0.0866	0.0375	2.31	0.0986	0.0427
1,800	2.81	0.138	0.0597	2.96	0.157	0.0680
2,200	3.43	0.200	0.0865	3.62	0.228	0.0985
2,600	4.06	0.272	0.118	4.28	0.310	0.134
3,000	4.68	0.355	0.154	4.94	0.404	0.175
3,500	5.46	0.472	0.204	5.76	0.537	0.233
4,000	6.24	0.604	0.261	6.59	0.688	0.298
4,500	7.03	0.751	0.325	7.41	0.855	0.370
5,000	7.81	0.912	0.395	8.24	1.04	0.450
5,500	8.59	1.09	0.471	9.06	1.24	0.537
6,000	9.37	1.28	0.553	9.88	1.46	0.630
6,500	10.1	1.48	0.642	10.7	1.69	0.731
7,000	10.9	1.70	0.736	11.5	1.94	0.838

Table B-2.D Flow friction loss, AWWA C900 (and ASTM D2241) 20-in. IPS OD PVC pipe

	20-in. IPS								
	DR 41 Pressure rated 100 psi, D_i = 19.17 in.			DR 32.5 Pressure rated 125 psi, D_i = 18.70 in.			DR 26 Pressure rated 160 psi, D_i = 18.37 in.		
		Pressure drop			Pressure drop			Pressure drop	
Flow, gpm	Velocity, ft/s	ft H_2O/100 ft	psi/100 ft	Velocity, ft/s	ft H_2O/100 ft	psi/100 ft	Velocity, ft/s	ft H_2O/100 ft	psi/100 ft
500	0.568	0.00596	0.00258	0.585	0.00639	0.00277	0.606	0.00696	0.00301
800	0.909	0.0142	0.00616	0.936	0.0152	0.00660	0.969	0.0166	0.00719
1,100	1.25	0.0256	0.0111	1.29	0.0275	0.0119	1.33	0.0299	0.0130
1,400	1.59	0.0400	0.0173	1.64	0.0429	0.0186	1.70	0.0468	0.0202
1,700	1.93	0.0573	0.0248	1.99	0.0615	0.0266	2.06	0.0670	0.0290
2,000	2.27	0.0774	0.0335	2.34	0.0830	0.0360	2.42	0.0905	0.0392
2,500	2.84	0.117	0.0507	2.92	0.125	0.0543	3.03	0.137	0.0592
3,000	3.41	0.164	0.0710	3.51	0.176	0.0761	3.64	0.192	0.0829
3,500	3.98	0.218	0.0944	4.09	0.234	0.101	4.24	0.255	0.110
4,000	4.55	0.279	0.121	4.68	0.299	0.130	4.85	0.326	0.141
4,500	5.11	0.347	0.150	5.26	0.372	0.161	5.45	0.405	0.176
5,000	5.68	0.422	0.183	5.85	0.452	0.196	6.06	0.493	0.213
5,500	6.25	0.503	0.218	6.43	0.540	0.234	6.66	0.588	0.254
6,000	6.82	0.591	0.256	7.02	0.634	0.274	7.27	0.690	0.299
6,500	7.39	0.685	0.297	7.60	0.735	0.318	7.88	0.801	0.347
7,000	7.96	0.786	0.340	8.19	0.843	0.365	8.48	0.918	0.398
7,500	8.52	0.893	0.387	8.77	0.958	0.415	9.09	1.04	0.452
8,000	9.09	1.01	0.436	9.36	1.08	0.467	9.69	1.18	0.509

Notes:
1. Table is based on Eq 3-11
2. Friction-loss values are based on average $D_i = D_o - (2 \times 106\%\ t_{min}) = D_o - (2.12 \times t_{min})$, where:
   D_i = pipe inside diameter, in.
   D_o = pipe outside diameter, in.
   t_{min} = minimum wall thickness, in.

Table B-2.D Flow friction loss, AWWA C900 (and ASTM D2241) 20-in. IPS OD PVC pipe *(continued)*

	20-in. IPS					
	DR 21 Pressure rated 200 psi, D_i = 17.98 in.			DR 17 Pressure rated 250 psi, D_i = 17.51 in.		
		Pressure drop			Pressure drop	
Flow, gpm	Velocity, ft/s	ft H_2O/100 ft	psi/100 ft	Velocity, ft/s	ft H_2O/100 ft	psi/100 ft
500	0.632	0.00772	0.00334	0.667	0.00879	0.00381
800	1.01	0.0184	0.00798	1.07	0.0210	0.00909
1,100	1.39	0.0332	0.0144	1.47	0.0378	0.0164
1,400	1.77	0.0519	0.0225	1.87	0.0591	0.0256
1,700	2.15	0.0743	0.0322	2.27	0.0846	0.0366
2,000	2.53	0.100	0.0435	2.67	0.114	0.0495
2,500	3.16	0.152	0.0657	3.34	0.173	0.0748
3,000	3.79	0.212	0.0920	4.00	0.242	0.105
3,500	4.43	0.283	0.122	4.67	0.322	0.139
4,000	5.06	0.362	0.157	5.34	0.412	0.178
4,500	5.69	0.450	0.195	6.00	0.512	0.222
5,000	6.32	0.547	0.237	6.67	0.623	0.270
5,500	6.96	0.652	0.282	7.34	0.743	0.322
6,000	7.59	0.766	0.332	8.00	0.872	0.378
6,500	8.22	0.888	0.385	8.67	1.01	0.438
7,000	8.85	1.02	0.441	9.34	1.16	0.502
7,500	9.48	1.16	0.501	10.0	1.32	0.571
8,000	10.1	1.30	0.565	10.7	1.49	0.643

Table B-2.E Flow friction loss, AWWA C900 (and ASTM D2241) 24-in. IPS OD PVC pipe

	24-in. IPS								
	DR 41 Pressure rated 100 psi, D_i = 23.00 in.			DR 32.5 Pressure rated 125 psi, D_i = 22.43 in.			DR 26 Pressure rated 160 psi, D_i = 22.04 in.		
		Pressure drop			Pressure drop			Pressure drop	
Flow, gpm	Velocity, ft/s	ft H_2O/100 ft	psi/100 ft	Velocity, ft/s	ft H_2O/100 ft	psi/100 ft	Velocity, ft/s	ft H_2O/100 ft	psi/100 ft
700	0.553	0.00458	0.00198	0.569	0.00491	0.00213	0.589	0.00535	0.00232
1,100	0.868	0.0106	0.00457	0.894	0.0113	0.00490	0.926	0.0123	0.00534
1,500	1.18	0.0187	0.00812	1.22	0.0201	0.00871	1.26	0.0219	0.00948
2,000	1.58	0.0319	0.0138	1.62	0.0342	0.0148	1.68	0.0373	0.0161
2,500	1.97	0.0482	0.0209	2.03	0.0517	0.0224	2.10	0.0563	0.0244
3,000	2.37	0.0676	0.0293	2.44	0.0725	0.0314	2.52	0.0790	0.0342
3,500	2.76	0.0899	0.0389	2.84	0.0964	0.0417	2.95	0.105	0.0455
4,000	3.16	0.115	0.0498	3.25	0.123	0.0534	3.37	0.134	0.0582
4,500	3.55	0.143	0.0620	3.66	0.153	0.0664	3.79	0.167	0.0724
5,000	3.95	0.174	0.0753	4.06	0.186	0.0807	4.21	0.203	0.0880
5,500	4.34	0.207	0.0898	4.47	0.222	0.0963	4.63	0.242	0.105
6,000	4.74	0.244	0.106	4.87	0.261	0.113	5.05	0.285	0.123
7,000	5.53	0.324	0.140	5.69	0.348	0.150	5.89	0.379	0.164
8,000	6.31	0.415	0.180	6.50	0.445	0.193	6.73	0.485	0.210
9,000	7.10	0.516	0.223	7.31	0.553	0.240	7.57	0.603	0.261
10,000	7.89	0.627	0.271	8.12	0.672	0.291	8.41	0.732	0.317
11,000	8.68	0.748	0.324	8.94	0.802	0.347	9.26	0.874	0.378
12,000	9.47	0.878	0.380	9.75	0.942	0.408	10.1	1.03	0.444

Notes:
1. Table is based on Eq 3-11
2. Friction-loss values are based on average $D_i = D_o - (2 \times 106\%\ t_{min}) = D_o - (2.12 \times t_{min})$, where:
   D_i = pipe inside diameter, in.
   D_o = pipe outside diameter, in.
   t_{min} = minimum wall thickness, in.

Table B-2.E Flow friction loss, AWWA C900 (and ASTM D2241) 24-in. IPS OD PVC pipe *(continued)*

24-in. IPS						
	DR 21 Pressure rated 200 psi, D_i = 21.58 in.			DR 17 Pressure rated 250 psi, D_i = 21.01 in.		
		Pressure drop			Pressure drop	
Flow, gpm	Velocity, ft/s	ft H_2O/100 ft	psi/100 ft	Velocity, ft/s	ft H_2O/100 ft	psi/100 ft
700	0.615	0.00593	0.00257	0.649	0.00676	0.00293
1,100	0.966	0.0137	0.00593	1.02	0.0156	0.00675
1,500	1.32	0.0243	0.0105	1.39	0.0277	0.0120
2,000	1.76	0.0414	0.0179	1.85	0.0471	0.0204
2,500	2.20	0.0625	0.0271	2.32	0.0712	0.0308
3,000	2.63	0.0876	0.0379	2.78	0.100	0.0432
3,500	3.07	0.116	0.0504	3.24	0.133	0.0575
4,000	3.51	0.149	0.0646	3.71	0.170	0.0736
4,500	3.95	0.185	0.0803	4.17	0.211	0.0915
5,000	4.39	0.225	0.0976	4.63	0.257	0.111
5,500	4.83	0.269	0.116	5.10	0.306	0.133
6,000	5.27	0.316	0.137	5.56	0.360	0.156
7,000	6.15	0.420	0.182	6.49	0.478	0.207
8,000	7.03	0.538	0.233	7.41	0.612	0.265
9,000	7.90	0.669	0.289	8.34	0.761	0.330
10,000	8.78	0.812	0.352	9.27	0.925	0.401
11,000	9.66	0.969	0.420	10.2	1.10	0.478
12,000	10.5	1.14	0.493	11.1	1.30	0.561

Table B-2.F Flow friction loss, AWWA C900 (and ASTM D2241) 30-in. IPS OD PVC pipe

	30-in. IPS								
	DR 41 Pressure rated 100 psi, D_i = 28.75 in.			DR 32.5 Pressure rated 125 psi, D_i = 28.04 in.			DR 26 Pressure rated 160 psi, D_i = 27.55 in.		
		Pressure drop			Pressure drop			Pressure drop	
Flow, gpm	Velocity, ft/s	ft H_2O/100 ft	psi/100 ft	Velocity, ft/s	ft H_2O/100 ft	psi/100 ft	Velocity, ft/s	ft H_2O/100 ft	psi/100 ft
1,000	0.505	0.00299	0.00130	0.520	0.00321	0.00139	0.539	0.00350	0.00151
1,500	0.758	0.00634	0.00274	0.780	0.00680	0.00294	0.808	0.00740	0.00321
2,000	1.01	0.0108	0.00467	1.04	0.0116	0.00501	1.08	0.0126	0.00546
2,500	1.26	0.0163	0.00706	1.30	0.0175	0.00757	1.35	0.0191	0.00825
3,000	1.52	0.0229	0.00989	1.56	0.0245	0.0106	1.62	0.0267	0.0116
3,800	1.92	0.0354	0.0153	1.98	0.0379	0.0164	2.05	0.0413	0.0179
4,600	2.32	0.0504	0.0218	2.39	0.0540	0.0234	2.48	0.0589	0.0255
5,400	2.73	0.0678	0.0294	2.81	0.0727	0.0315	2.91	0.0792	0.0343
6,200	3.13	0.0875	0.0379	3.22	0.0939	0.0406	3.34	0.102	0.0443
7,000	3.54	0.110	0.0474	3.64	0.117	0.0509	3.77	0.128	0.0554
8,500	4.29	0.157	0.0679	4.42	0.168	0.0729	4.58	0.183	0.0794
10,000	5.05	0.212	0.0918	5.20	0.227	0.0984	5.39	0.248	0.107
11,500	5.81	0.274	0.119	5.98	0.294	0.127	6.19	0.321	0.139
13,000	6.57	0.344	0.149	6.76	0.369	0.160	7.00	0.402	0.174
14,500	7.33	0.421	0.183	7.54	0.452	0.196	7.81	0.492	0.213
16,000	8.08	0.506	0.219	8.32	0.542	0.235	8.62	0.591	0.256
18,000	9.09	0.629	0.272	9.36	0.674	0.292	9.69	0.734	0.318
20,000	10.1	0.764	0.331	10.4	0.819	0.355	10.8	0.893	0.386

Notes:
1. Table is based on Eq 3-11
2. Friction-loss values are based on average $D_i = D_o - (2 \times 106\%\ t_{min}) = D_o - (2.12 \times t_{min})$, where:
   D_i = pipe inside diameter, in.
   D_o = pipe outside diameter, in.
   t_{min} = minimum wall thickness, in.

Table B-2.F Flow friction loss, AWWA C900 (and ASTM D2241) 30-in. IPS OD PVC pipe *(continued)*

30-in. IPS						
	DR 21 Pressure rated 200 psi, D_i = 26.97 in.			DR 17 Pressure rated 250 psi, D_i = 26.26 in.		
		Pressure drop			Pressure drop	
Flow, gpm	Velocity, ft/s	ft H_2O/100 ft	psi/100 ft	Velocity, ft/s	ft H_2O/100 ft	psi/100 ft
1,000	0.562	0.00388	0.00168	0.593	0.00442	0.00191
1,500	0.843	0.00821	0.00356	0.889	0.00936	0.00405
2,000	1.12	0.0140	0.00605	1.19	0.0159	0.00690
2,500	1.40	0.0211	0.00915	1.48	0.0241	0.0104
3,000	1.69	0.0296	0.0128	1.78	0.0337	0.0146
3,800	2.14	0.0458	0.0199	2.25	0.0522	0.0226
4,600	2.59	0.0653	0.0283	2.73	0.0744	0.0322
5,400	3.03	0.0878	0.0380	3.20	0.100	0.0433
6,200	3.48	0.113	0.0491	3.68	0.129	0.0559
7,000	3.93	0.142	0.0615	4.15	0.162	0.0700
8,500	4.78	0.203	0.0880	5.04	0.232	0.100
10,000	5.62	0.275	0.119	5.93	0.313	0.135
11,500	6.46	0.356	0.154	6.82	0.405	0.175
13,000	7.31	0.446	0.193	7.71	0.508	0.220
14,500	8.15	0.546	0.236	8.60	0.622	0.269
16,000	8.99	0.655	0.284	9.49	0.746	0.323
18,000	10.1	0.815	0.353	10.7	0.928	0.402
20,000	11.2	0.990	0.429	11.9	1.13	0.488

Table B-2.G Flow friction loss, AWWA C900 (and ASTM D2241) 36-in. IPS OD PVC pipe

	36-in. IPS								
	DR 41 Pressure rated 100 psi, D_i = 34.50 in.			DR 32.5 Pressure rated 125 psi, D_i = 33.65 in.			DR 26 Pressure rated 160 psi, D_i = 33.06 in.		
		Pressure drop			Pressure drop			Pressure drop	
Flow, gpm	Velocity, ft/s	ft H_2O/100 ft	psi/100 ft	Velocity, ft/s	ft H_2O/100 ft	psi/100 ft	Velocity, ft/s	ft H_2O/100 ft	psi/100 ft
1,500	0.526	0.00261	0.00113	0.542	0.00280	0.00121	0.561	0.00305	0.00132
2,000	0.702	0.00445	0.00193	0.722	0.00477	0.00207	0.748	0.00520	0.00225
2,500	0.877	0.00672	0.00291	0.903	0.00721	0.00312	0.935	0.00785	0.00340
3,000	1.05	0.00942	0.00408	1.08	0.0101	0.00437	1.12	0.0110	0.00476
4,000	1.40	0.0160	0.00695	1.44	0.0172	0.00745	1.50	0.0187	0.00811
5,000	1.75	0.0242	0.0105	1.81	0.0260	0.0113	1.87	0.0283	0.0123
6,000	2.10	0.0340	0.0147	2.17	0.0364	0.0158	2.24	0.0397	0.0172
7,000	2.46	0.0452	0.0196	2.53	0.0484	0.0210	2.62	0.0528	0.0228
8,000	2.81	0.0578	0.0250	2.89	0.0620	0.0269	2.99	0.0675	0.0292
9,000	3.16	0.0719	0.0311	3.25	0.0771	0.0334	3.37	0.0840	0.0364
10,000	3.51	0.0874	0.0378	3.61	0.0937	0.0406	3.74	0.102	0.0442
12,000	4.21	0.122	0.0530	4.33	0.131	0.0568	4.49	0.143	0.0619
14,000	4.91	0.163	0.0705	5.05	0.175	0.0756	5.24	0.190	0.0824
16,000	5.61	0.208	0.0903	5.78	0.224	0.0968	5.98	0.244	0.105
18,000	6.31	0.259	0.112	6.50	0.278	0.120	6.73	0.303	0.131
20,000	7.02	0.315	0.136	7.22	0.338	0.146	7.48	0.368	0.159
24,000	8.42	0.441	0.191	8.67	0.473	0.205	8.98	0.516	0.223
28,000	9.82	0.587	0.254	10.1	0.629	0.273	10.5	0.686	0.297

Notes:
1. Table is based on Eq 3-11
2. Friction-loss values are based on average $D_i = D_o - (2 \times 106\%\ t_{min}) = D_o - (2.12 \times t_{min})$, where:
   D_i = pipe inside diameter, in.
   D_o = pipe outside diameter, in.
   t_{min} = minimum wall thickness, in.

Table B-2.G Flow friction loss, AWWA C900 (and ASTM D2241) 36-in. IPS OD PVC pipe *(continued)*

36-in. IPS						
	DR 21 Pressure rated 200 psi, D_i = 32.37 in.			DR 17 Pressure rated 250 psi, D_i = 7.55 in.		
		Pressure drop			Pressure drop	
Flow, gpm	Velocity, ft/s	ft H_2O/100 ft	psi/100 ft	Velocity, ft/s	ft H_2O/100 ft	psi/100 ft
1,500	0.585	0.00339	0.00147	0.618	0.00386	0.00167
2,000	0.781	0.00577	0.00250	0.824	0.00657	0.00284
2,500	0.976	0.00871	0.00377	1.03	0.00992	0.00430
3,000	1.17	0.0122	0.00529	1.24	0.0139	0.00602
4,000	1.56	0.0208	0.00900	1.65	0.0237	0.0103
5,000	1.95	0.0314	0.0136	2.06	0.0358	0.0155
6,000	2.34	0.0440	0.0191	2.47	0.0501	0.0217
7,000	2.73	0.0585	0.0253	2.88	0.0667	0.0289
8,000	3.12	0.0749	0.0324	3.29	0.0854	0.0370
9,000	3.51	0.0932	0.0403	3.71	0.106	0.0460
10,000	3.90	0.113	0.0490	4.12	0.129	0.0558
12,000	4.68	0.159	0.0687	4.94	0.181	0.0782
14,000	5.46	0.211	0.0914	5.76	0.240	0.104
16,000	6.24	0.270	0.117	6.59	0.308	0.133
18,000	7.03	0.336	0.145	7.41	0.383	0.166
20,000	7.81	0.408	0.177	8.24	0.465	0.201
24,000	9.37	0.572	0.248	9.88	0.651	0.282
28,000	10.9	0.761	0.329	11.5	0.866	0.375

Table B-3.A Flow friction loss, AWWA C909 (and CSA B137.3.1) 4-in. CIOD PVCO pipe

	4-in. CIOD								
	Pressure rated 165 Psi D_i = 4.57 in.			Pressure rated 235 psi D_i = 4.47 in.			Pressure rated 305 psi D_i = 4.38 in.		
Flow	Velocity	Pressure drop		Velocity	Pressure drop		Velocity	Pressure drop	
gpm	ft/s	ft H_2O/ 100 ft	psi/100 ft	ft/s	ft H_2O/ 100 ft	psi/100 ft	ft/s	ft H_2O/ 100 ft	psi/100 ft
25	0.490	0.0236	0.0102	0.511	0.0261	0.0113	0.533	0.0289	0.0125
40	0.783	0.0562	0.0244	0.817	0.0623	0.0270	0.852	0.0690	0.0299
55	1.08	0.101	0.0439	1.12	0.112	0.0486	1.17	0.124	0.0539
70	1.37	0.158	0.0686	1.43	0.175	0.0760	1.49	0.194	0.0842
85	1.66	0.227	0.0982	1.74	0.251	0.109	1.81	0.278	0.121
100	1.96	0.306	0.133	2.04	0.339	0.147	2.13	0.376	0.163
120	2.35	0.429	0.186	2.45	0.476	0.206	2.56	0.527	0.228
140	2.74	0.571	0.247	2.86	0.633	0.274	2.98	0.701	0.303
160	3.13	0.731	0.317	3.27	0.810	0.351	3.41	0.897	0.389
180	3.53	0.909	0.394	3.68	1.01	0.436	3.84	1.12	0.483
200	3.92	1.10	0.478	4.09	1.22	0.530	4.26	1.36	0.587
225	4.41	1.37	0.595	4.60	1.52	0.659	4.79	1.69	0.730
250	4.90	1.67	0.723	5.11	1.85	0.801	5.33	2.05	0.887
275	5.39	1.99	0.862	5.62	2.21	0.955	5.86	2.44	1.06
300	5.88	2.34	1.01	6.13	2.59	1.12	6.39	2.87	1.24
350	6.86	3.11	1.35	7.15	3.45	1.49	7.46	3.82	1.65
400	7.83	3.98	1.72	8.17	4.41	1.91	8.52	4.89	2.12
450	8.81	4.95	2.14	9.19	5.49	2.38	9.59	6.08	2.63

Notes:

1. CSA values were converted from metric units.
2. Table is based on Eq 3-11.
3. Friction-loss values are based on average $D_i = D_o - (2 \times 106\% \times t_{min}) = D_o - (2.12 \times t_{min})$ where:

D_i = pipe inside diameter, in.
D_o = pipe outside diameter, in.
t_{min} = minimum wall thickness, in.

Table B-3.B Flow friction loss, AWWA C909 (and CSA B137.3.1) 6-in. CIOD PVCO pipe

6-in. CIOD									
	Pressure rated 165 Psi D_i = 6.57 in.			Pressure rated 235 psi D_i = 6.43 in.			Pressure rated 305 psi D_i = 6.30 in.		
Flow	Velocity	Pressure drop		Velocity	Pressure drop		Velocity	Pressure drop	
gpm	ft/s	ft H_2O/ 100 ft	psi/100 ft	ft/s	ft H_2O/ 100 ft	psi/100 ft	ft/s	ft H_2O/ 100 ft	psi/100 ft
50	0.474	0.0146	0.00631	0.494	0.0161	0.00699	0.515	0.0179	0.00774
75	0.711	0.0309	0.0134	0.741	0.0342	0.0148	0.773	0.0378	0.0164
100	0.948	0.0525	0.0228	0.988	0.0582	0.0252	1.03	0.0644	0.0279
125	1.19	0.0794	0.0344	1.24	0.0879	0.0381	1.29	0.0973	0.0421
150	1.42	0.111	0.0482	1.48	0.123	0.0533	1.55	0.136	0.0590
175	1.66	0.148	0.0641	1.73	0.164	0.0709	1.80	0.181	0.0785
200	1.90	0.189	0.0820	1.98	0.210	0.0908	2.06	0.232	0.101
250	2.37	0.286	0.124	2.47	0.317	0.137	2.58	0.351	0.152
300	2.84	0.401	0.174	2.97	0.444	0.192	3.09	0.492	0.213
350	3.32	0.533	0.231	3.46	0.590	0.256	3.61	0.654	0.283
400	3.79	0.683	0.296	3.95	0.756	0.327	4.12	0.837	0.362
450	4.27	0.849	0.368	4.45	0.940	0.407	4.64	1.04	0.451
500	4.74	1.03	0.447	4.94	1.14	0.495	5.15	1.26	0.548
600	5.69	1.45	0.626	5.93	1.60	0.693	6.18	1.77	0.767
700	6.64	1.92	0.833	6.92	2.13	0.922	7.22	2.36	1.02
800	7.58	2.46	1.07	7.91	2.72	1.18	8.25	3.02	1.31
900	8.53	3.06	1.33	8.90	3.39	1.47	9.28	3.75	1.62
1,000	9.48	3.72	1.61	9.88	4.12	1.78	10.3	4.56	1.97

Table B-3.C Flow friction loss, AWWA C909 (and CSA B137.3.1) 8-in. CIOD PVCO pipe

	8-in. CIOD								
	Pressure rated 165 Psi D_i = 8.61 in.			Pressure rated 235 psi D_i = 8.44 in.			Pressure rated 305 psi D_i = 8.26 in.		
Flow	Velocity	Pressure drop		Velocity	Pressure drop		Velocity	Pressure drop	
gpm	ft/s	ft H_2O/ 100 ft	psi/100 ft	ft/s	ft H_2O/ 100 ft	psi/100 ft	ft/s	ft H_2O/ 100 ft	psi/100 ft
100	0.551	0.0141	0.00609	0.575	0.0156	0.00674	0.599	0.0172	0.00747
125	0.689	0.0213	0.00920	0.718	0.0235	0.0102	0.749	0.0261	0.0113
150	0.827	0.0298	0.0129	0.862	0.0330	0.0143	0.899	0.0365	0.0158
200	1.10	0.0507	0.0220	1.15	0.0561	0.0243	1.20	0.0622	0.0269
250	1.38	0.0766	0.0332	1.44	0.0848	0.0367	1.50	0.0939	0.0407
300	1.65	0.107	0.0465	1.72	0.119	0.0514	1.80	0.132	0.0570
350	1.93	0.143	0.0618	2.01	0.158	0.0684	2.10	0.175	0.0758
400	2.20	0.183	0.0791	2.30	0.202	0.0876	2.40	0.224	0.0970
450	2.48	0.227	0.0984	2.59	0.252	0.109	2.70	0.279	0.121
500	2.76	0.276	0.120	2.87	0.306	0.132	3.00	0.339	0.147
600	3.31	0.387	0.168	3.45	0.428	0.185	3.60	0.475	0.205
700	3.86	0.515	0.223	4.02	0.570	0.247	4.20	0.631	0.273
800	4.41	0.659	0.285	4.60	0.729	0.316	4.79	0.808	0.350
1,000	5.51	1.00	0.431	5.75	1.10	0.477	5.99	1.22	0.529
1,200	6.61	1.39	0.604	6.90	1.54	0.669	7.19	1.71	0.741
1,400	7.72	1.86	0.803	8.04	2.05	0.889	8.39	2.28	0.985
1,600	8.82	2.38	1.03	9.19	2.63	1.14	9.59	2.91	1.26
1,800	9.92	2.95	1.28	10.3	3.27	1.42	10.8	3.62	1.57

Notes:
1. CSA values were converted from metric units.
2. Table is based on Eq 3-11.
3. Friction-loss values are based on average $D_i = D_o - (2 \times 106\% \times t_{min}) = D_o - (2.12 \times t_{min})$ where:
   D_i = pipe inside diameter, in.
   D_o = pipe outside diameter, in.
   t_{min} = minimum wall thickness, in.

Table B-3.D Flow friction loss, AWWA C909 (and CSA B137.3.1) 10-in. CIOD PVCO pipe

10-in. CIOD									
	Pressure rated 165 Psi D_i = 10.57 in.			Pressure rated 235 psi D_i = 10.35 in.			Pressure rated 305 psi D_i = 10.13 in.		
Flow	Velocity	Pressure drop		Velocity	Pressure drop		Velocity	Pressure drop	
gpm	ft/s	ft H_2O/ 100 ft	psi/100 ft	ft/s	ft H_2O/ 100 ft	psi/100 ft	ft/s	ft H_2O/ 100 ft	psi/100 ft
150	0.549	0.0110	0.00478	0.573	0.0122	0.00529	0.598	0.0135	0.00586
200	0.732	0.0188	0.00813	0.764	0.0208	0.00901	0.797	0.0230	0.0100
250	0.916	0.0284	0.0123	0.955	0.0314	0.0136	1.00	0.0348	0.0151
300	1.10	0.0398	0.0172	1.15	0.0441	0.0191	1.20	0.0488	0.0211
350	1.28	0.0529	0.0229	1.34	0.0586	0.0254	1.39	0.0649	0.0281
400	1.46	0.0677	0.0293	1.53	0.0750	0.0325	1.59	0.0830	0.0360
450	1.65	0.0842	0.0365	1.72	0.0933	0.0404	1.79	0.103	0.0447
500	1.83	0.102	0.0443	1.91	0.113	0.0491	1.99	0.125	0.0543
600	2.20	0.143	0.0621	2.29	0.159	0.0688	2.39	0.176	0.0761
700	2.56	0.191	0.0826	2.67	0.211	0.0915	2.79	0.234	0.101
800	2.93	0.244	0.106	3.06	0.270	0.117	3.19	0.299	0.130
1,000	3.66	0.369	0.160	3.82	0.409	0.177	3.98	0.452	0.196
1,200	4.39	0.517	0.224	4.58	0.573	0.248	4.78	0.634	0.274
1,400	5.13	0.687	0.298	5.35	0.762	0.330	5.58	0.843	0.365
1,600	5.86	0.880	0.381	6.11	0.975	0.422	6.37	1.08	0.467
1,800	6.59	1.09	0.474	6.88	1.21	0.525	7.17	1.34	0.581
2,200	8.06	1.59	0.687	8.40	1.76	0.761	8.76	1.95	0.842
2,600	9.52	2.16	0.935	9.93	2.39	1.04	10.4	2.65	1.15

Table B-3.E Flow friction loss, AWWA C909 (and CSA B137.3.1) 12-in. CIOD PVCO pipe

	12-in. CIOD								
	Pressure rated 165 Psi $D_i = 12.56$ in.			Pressure rated 235 psi $D_i = 12.30$ in.			Pressure rated 305 psi $D_i = 12.05$ in.		
		Pressure drop			Pressure drop			Pressure drop	
Flow gpm	Velocity ft/s	ft H_2O/ 100 ft	psi/100 ft	Velocity ft/s	ft H_2O/ 100 ft	psi/100 ft	Velocity ft/s	ft H_2O/ 100 ft	psi/100 ft
200	0.518	0.00809	0.00350	0.540	0.00896	0.00388	0.563	0.00993	0.00430
250	0.648	0.0122	0.00530	0.675	0.0135	0.00586	0.704	0.0150	0.00650
300	0.777	0.0171	0.00742	0.810	0.0190	0.00822	0.845	0.0210	0.00910
350	0.907	0.0228	0.00987	0.945	0.0252	0.0109	0.986	0.0280	0.0121
400	1.04	0.0292	0.0126	1.08	0.0323	0.0140	1.13	0.0358	0.0155
500	1.30	0.0441	0.0191	1.35	0.0488	0.0211	1.41	0.0541	0.0234
600	1.55	0.0618	0.0267	1.62	0.0684	0.0296	1.69	0.0758	0.0328
700	1.81	0.0822	0.0356	1.89	0.0910	0.0394	1.97	0.101	0.0436
800	2.07	0.105	0.0455	2.16	0.116	0.0504	2.25	0.129	0.0559
1,000	2.59	0.159	0.0688	2.70	0.176	0.0762	2.82	0.195	0.0844
1,200	3.11	0.223	0.0964	3.24	0.247	0.107	3.38	0.273	0.118
1,400	3.63	0.296	0.128	3.78	0.328	0.142	3.94	0.363	0.157
1,600	4.14	0.379	0.164	4.32	0.420	0.182	4.51	0.465	0.201
2,000	5.18	0.573	0.248	5.40	0.634	0.275	5.63	0.703	0.304
2,400	6.22	0.803	0.348	6.48	0.889	0.385	6.76	0.985	0.426
2,800	7.25	1.07	0.462	7.56	1.18	0.512	7.89	1.31	0.567
3,200	8.29	1.37	0.592	8.64	1.51	0.655	9.02	1.68	0.726
3,600	9.32	1.70	0.736	9.72	1.88	0.815	10.1	2.09	0.903

Notes:
1. CSA values were converted from metric units.
2. Table is based on Eq 3-11.
3. Friction-loss values are based on average $D_i = D_o - (2 \times 106\% \times t_{min}) = D_o - (2.12 \times t_{min})$ where:
   - D_i = pipe inside diameter, in.
   - D_o = pipe outside diameter, in.
   - t_{min} = minimum wall thickness, in.

Table B-3.F Flow friction loss, AWWA C909 (and CSA B137.3.1) 14-in. CIOD PVCO pipe

14-in. CIOD									
	Pressure rated 165 Psi D_i = 14.56 in.			Pressure rated 235 psi D_i = 14.26 in.			Pressure rated 305 psi D_i = 13.96 in.		
		Pressure drop			Pressure drop			Pressure drop	
Flow gpm	Velocity ft/s	ft H_2O/ 100 ft	psi/100 ft	Velocity ft/s	ft H_2O/ 100 ft	psi/100 ft	Velocity ft/s	ft H_2O/ 100 ft	psi/100 ft
300	0.578	0.00836	0.00362	0.603	0.00926	0.00401	0.629	0.0103	0.00444
450	0.867	0.0177	0.00766	0.905	0.0196	0.00849	0.944	0.0217	0.00940
600	1.16	0.0301	0.0130	1.21	0.0334	0.0144	1.26	0.0370	0.0160
800	1.54	0.0513	0.0222	1.61	0.0568	0.0246	1.68	0.0629	0.0273
1,000	1.93	0.0775	0.0336	2.01	0.0859	0.0372	2.10	0.0951	0.0412
1,200	2.31	0.109	0.0470	2.41	0.120	0.0521	2.52	0.133	0.0577
1,400	2.70	0.144	0.0625	2.81	0.160	0.0693	2.94	0.177	0.0767
1,600	3.08	0.185	0.0801	3.22	0.205	0.0887	3.35	0.227	0.0982
1,800	3.47	0.230	0.100	3.62	0.255	0.110	3.77	0.282	0.122
2,000	3.86	0.279	0.121	4.02	0.310	0.134	4.19	0.343	0.148
2,200	4.24	0.333	0.144	4.42	0.369	0.160	4.61	0.409	0.177
2,600	5.01	0.454	0.197	5.23	0.503	0.218	5.45	0.557	0.241
3,000	5.78	0.592	0.256	6.03	0.655	0.284	6.29	0.726	0.314
3,400	6.55	0.746	0.323	6.84	0.826	0.358	7.13	0.915	0.396
3,800	7.32	0.916	0.397	7.64	1.01	0.439	7.97	1.12	0.487
4,200	8.10	1.10	0.477	8.44	1.22	0.529	8.81	1.35	0.586
4,600	8.87	1.30	0.565	9.25	1.45	0.626	9.64	1.60	0.693
5,000	9.64	1.52	0.659	10.1	1.69	0.730	10.5	1.87	0.809

Table B-3.G Flow friction loss, AWWA C909 (and CSA B137.3.1) 16-in. CIOD PVCO pipe

	16-in. CIOD								
	Pressure rated 165 Psi D_i = 16.56 in.			Pressure rated 235 psi D_i = 16.22 in.			Pressure rated 305 psi D_i = 15.88 in.		
Flow	Velocity	Pressure drop		Velocity	Pressure drop		Velocity	Pressure drop	
gpm	ft/s	ft H_2O/ 100 ft	psi/100 ft	ft/s	ft H_2O/ 100 ft	psi/100 ft	ft/s	ft H_2O/ 100 ft	psi/100 ft
350	0.522	0.00595	0.00258	0.544	0.00659	0.00285	0.567	0.00730	0.00316
500	0.745	0.0115	0.00498	0.777	0.0127	0.00552	0.811	0.0141	0.00612
650	0.969	0.0187	0.00810	1.01	0.0207	0.00897	1.05	0.0229	0.00994
800	1.19	0.0275	0.0119	1.24	0.0304	0.0132	1.30	0.0337	0.0146
1,000	1.49	0.0415	0.0180	1.55	0.0459	0.0199	1.62	0.0509	0.0220
1,200	1.79	0.0581	0.0252	1.87	0.0644	0.0279	1.95	0.0713	0.0309
1,400	2.09	0.0773	0.0335	2.18	0.0856	0.0371	2.27	0.0949	0.0411
1,800	2.68	0.123	0.0533	2.80	0.136	0.0590	2.92	0.151	0.0654
2,200	3.28	0.178	0.0773	3.42	0.198	0.0856	3.57	0.219	0.0948
2,600	3.88	0.243	0.105	4.04	0.269	0.117	4.22	0.298	0.129
3,000	4.47	0.317	0.137	4.66	0.351	0.152	4.86	0.389	0.168
3,500	5.22	0.421	0.182	5.44	0.466	0.202	5.67	0.517	0.224
4,000	5.96	0.539	0.234	6.22	0.597	0.259	6.49	0.662	0.287
4,500	6.71	0.671	0.290	6.99	0.743	0.322	7.30	0.823	0.356
5,000	7.45	0.815	0.353	7.77	0.902	0.391	8.11	1.00	0.433
5,500	8.20	0.972	0.421	8.55	1.08	0.466	8.92	1.19	0.516
6,000	8.94	1.14	0.494	9.33	1.26	0.547	9.73	1.40	0.607
6,500	9.69	1.32	0.573	10.1	1.47	0.635	10.5	1.62	0.703

Notes:
1. CSA values were converted from metric units.
2. Table is based on Eq 3-11.
3. Friction-loss values are based on average $D_i = D_o - (2 \times 106\% \times t_{min}) = D_o - (2.12 \times t_{min})$ where:
   D_i = pipe inside diameter, in.
   D_o = pipe outside diameter, in.
   t_{min} = minimum wall thickness, in.

Table B-3.H Flow friction loss, AWWA C909 (and CSA B137.3.1) 18-in. CIOD PVCO pipe

18-in. CIOD									
	Pressure rated 165 Psi D_i = 18.56 in.			Pressure rated 235 psi D_i = 18.18 in.			Pressure rated 305 psi D_i = 17.80 in.		
Flow gpm	Velocity ft/s	Pressure drop		Velocity ft/s	Pressure drop		Velocity ft/s	Pressure drop	
		ft H_2O/ 100 ft	psi/100 ft		ft H_2O/ 100 ft	psi/100 ft		ft H_2O/ 100 ft	psi/100 ft
500	0.593	0.00662	0.00287	0.619	0.00733	0.00317	0.645	0.00812	0.00351
800	0.949	0.0158	0.00684	0.990	0.0175	0.00757	1.03	0.0194	0.00838
1,100	1.31	0.0285	0.0123	1.36	0.0315	0.0136	1.42	0.0349	0.0151
1,400	1.66	0.0445	0.0193	1.73	0.0492	0.0213	1.81	0.0545	0.0236
1,800	2.14	0.0708	0.0306	2.23	0.0784	0.0339	2.32	0.0868	0.0376
2,200	2.61	0.103	0.0444	2.72	0.114	0.0492	2.84	0.126	0.0545
2,600	3.09	0.140	0.0605	3.22	0.155	0.0670	3.36	0.171	0.0742
3,000	3.56	0.182	0.0788	3.71	0.202	0.0873	3.87	0.223	0.0967
3,500	4.15	0.242	0.105	4.33	0.268	0.116	4.52	0.297	0.129
4,000	4.75	0.310	0.134	4.95	0.343	0.149	5.16	0.380	0.165
4,500	5.34	0.386	0.167	5.57	0.427	0.185	5.81	0.473	0.205
5,000	5.93	0.468	0.203	6.19	0.519	0.225	6.45	0.575	0.249
5,500	6.53	0.559	0.242	6.81	0.619	0.268	7.10	0.685	0.297
6,000	7.12	0.656	0.284	7.43	0.727	0.315	7.74	0.805	0.349
6,500	7.71	0.761	0.330	8.05	0.843	0.365	8.39	0.934	0.404
7,000	8.31	0.873	0.378	8.66	0.967	0.419	9.04	1.07	0.464
7,500	8.90	0.992	0.429	9.28	1.10	0.476	9.68	1.22	0.527
8,000	9.49	1.12	0.484	9.90	1.24	0.536	10.3	1.37	0.594

Table B-3.I Flow friction loss, AWWA C909 (and CSA B137.3.1) 20-in. CIOD PVCO pipe

20-in. CIOD									
	Pressure rated 165 Psi D_i = 20.56 in.			Pressure rated 235 psi D_i = 20.13 in.			Pressure rated 305 psi D_i = 19.71 in.		
Flow gpm	Velocity ft/s	Pressure drop ft H_2O/ 100 ft	Pressure drop psi/100 ft	Velocity ft/s	Pressure drop ft H_2O/ 100 ft	Pressure drop psi/100 ft	Velocity ft/s	Pressure drop ft H_2O/ 100 ft	Pressure drop psi/100 ft
600	0.580	0.00564	0.00244	0.605	0.00625	0.00270	0.631	0.00692	0.00300
1,000	0.967	0.0145	0.00628	1.01	0.0161	0.00696	1.05	0.0178	0.00771
1,400	1.35	0.0270	0.0117	1.41	0.0299	0.0130	1.47	0.0332	0.0144
1,800	1.74	0.0431	0.0186	1.82	0.0477	0.0206	1.89	0.0528	0.0229
2,200	2.13	0.0624	0.0270	2.22	0.0691	0.0299	2.31	0.0766	0.0331
2,600	2.52	0.0850	0.0368	2.62	0.0941	0.0408	2.74	0.104	0.0452
3,000	2.90	0.111	0.0480	3.03	0.123	0.0531	3.16	0.136	0.0588
3,500	3.39	0.147	0.0638	3.53	0.163	0.0706	3.68	0.181	0.0783
4,000	3.87	0.189	0.0817	4.03	0.209	0.0904	4.21	0.231	0.100
4,500	4.35	0.235	0.102	4.54	0.260	0.112	4.73	0.288	0.125
5,000	4.84	0.285	0.123	5.04	0.316	0.137	5.26	0.350	0.151
5,500	5.32	0.340	0.147	5.55	0.376	0.163	5.79	0.417	0.181
6,000	5.80	0.399	0.173	6.05	0.442	0.191	6.31	0.490	0.212
6,500	6.29	0.463	0.201	6.56	0.513	0.222	6.84	0.568	0.246
7,000	6.77	0.531	0.230	7.06	0.588	0.255	7.36	0.651	0.282
8,000	7.74	0.680	0.294	8.07	0.753	0.326	8.42	0.834	0.361
9,000	8.71	0.846	0.366	9.08	0.936	0.405	9.47	1.04	0.449
10,000	9.67	1.03	0.445	10.1	1.14	0.493	10.5	1.26	0.546

Notes:

1. CSA values were converted from metric units.
2. Table is based on Eq 3-11.
3. Friction-loss values are based on average $D_i = D_o - (2 \times 106\% \times t_{min}) = D_o - (2.12 \times t_{min})$ where:
   D_i = pipe inside diameter, in.
   D_o = pipe outside diameter, in.
   t_{min} = minimum wall thickness, in.

Table B-3.J Flow friction loss, AWWA C909 (and CSA B137.3.1) 24-in. CIOD PVCO pipe

24-in. CIOD									
	Pressure rated 165 Psi D_i = 24.56 in.			Pressure rated 235 psi D_i = 24.05 in.			Pressure rated 305 psi D_i = 23.55 in.		
	Velocity	Pressure drop		Velocity	Pressure drop		Velocity	Pressure drop	
Flow gpm	ft/s	ft H_2O/ 100 ft	psi/100 ft	ft/s	ft H_2O/ 100 ft	psi/100 ft	ft/s	ft H_2O/ 100 ft	psi/100 ft
800	0.542	0.00405	0.00175	0.566	0.00449	0.00194	0.590	0.00497	0.00215
1,200	0.814	0.00857	0.00371	0.848	0.00950	0.00411	0.885	0.0105	0.00455
1,600	1.08	0.0146	0.00632	1.13	0.0162	0.00700	1.18	0.0179	0.00775
2,000	1.36	0.0221	0.00955	1.41	0.0244	0.0106	1.47	0.0271	0.0117
2,600	1.76	0.0358	0.0155	1.84	0.0397	0.0172	1.92	0.0440	0.0190
3,200	2.17	0.0526	0.0228	2.26	0.0583	0.0252	2.36	0.0646	0.0280
3,800	2.58	0.0723	0.0313	2.69	0.0801	0.0347	2.80	0.0887	0.0384
4,600	3.12	0.103	0.0446	3.25	0.114	0.0494	3.39	0.126	0.0547
5,400	3.66	0.139	0.0600	3.82	0.153	0.0664	3.98	0.170	0.0736
6,200	4.20	0.179	0.0775	4.38	0.198	0.0858	4.57	0.219	0.0950
7,000	4.75	0.224	0.0970	4.95	0.248	0.107	5.16	0.275	0.119
8,000	5.42	0.287	0.124	5.66	0.318	0.137	5.90	0.352	0.152
9,000	6.10	0.356	0.154	6.36	0.395	0.171	6.64	0.437	0.189
10,000	6.78	0.433	0.188	7.07	0.480	0.208	7.37	0.531	0.230
11,000	7.46	0.517	0.224	7.78	0.572	0.248	8.11	0.634	0.274
12,000	8.14	0.607	0.263	8.48	0.672	0.291	8.85	0.745	0.322
13,500	9.15	0.755	0.327	9.55	0.836	0.362	9.96	0.926	0.401
15,000	10.2	0.917	0.397	10.6	1.02	0.440	11.1	1.13	0.487

This page intentionally blank.

Index

Note: *f.* indicates figure; *t.* indicates table

Q

R

S

T

U

V

W

AWWA Manuals

M1, *Principles of Water Rates, Fees, and Charges*, #30001
M2, *Instrumentation and Control*, #30002
M3, *Safety Management for Water Utilities*, #30003
M4, *Water Fluoridation Principles and Practices*, #30004
M5, *Water Utility Management*, #30005
M6, *Water Meters—Selection, Installation, Testing, and Maintenance*, #30006
M7, *Problem Organisms in Water: Identification and Treatment*, #30007
M9, *Concrete Pressure Pipe*, #30009
M11, *Steel Pipe—A Guide for Design and Installation*, #30011
M12, *Simplified Procedures for Water Examination*, #30012
M14, *Backflow Prevention and Cross-Connection Control: Recommended Practices*, #30014
M17, *Fire Hydrants: Installation, Field Testing, and Maintenance*, #30017
M19, *Emergency Planning for Water and Wastewater Utilities*, #30019
M20, *Water Chlorination/Chloramination Practices and Principles*, #30020
M21, *Groundwater*, #30021
M22, *Sizing Water Service Lines and Meters*, #30022
M23, *PVC Pipe—Design and Installation*, #30023
M24, *Planning for the Distribution of Reclaimed Water*, #30024
M25, *Flexible-Membrane Covers and Linings for Potable-Water Reservoirs*, #30025
M27, *External Corrosion Control for Infrastructure Sustainability*, #30027
M28, *Rehabilitation of Water Mains*, #30028
M29, *Water Utility Capital Financing*, #30029
M30, *Precoat Filtration*, #30030
M31, *Distribution System Requirements for Fire Protection*, #30031
M32, *Computer Modeling of Water Distribution Systems*, #30032
M33, *Flowmeters in Water Supply*, #30033
M36, *Water Audits and Loss Control Programs*, #30036
M37, *Operational Control of Coagulation and Filtration Processes*, #30037
M38, *Electrodialysis and Electrodialysis Reversal*, #30038
M41, *Ductile-Iron Pipe and Fittings*, #30041
M42, *Steel Water-Storage Tanks*, #30042
M44, *Distribution Valves: Selection, Installation, Field Testing, and Maintenance*, #30044
M45, *Fiberglass Pipe Design*, #30045
M46, *Reverse Osmosis and Nanofiltration*, #30046
M47, *Capital Project Delivery*, #30047
M48, *Waterborne Pathogens*, #30048
M49, *Quarter-Turn Valves: Head Loss, Torque, and Cavitation Analysis*, #30049
M50, *Water Resources Planning*, #30050
M51, *Air Valves: Air-Release, Air/Vacuum and Combination*, #30051
M52, *Water Conservation Programs—A Planning Manual*, #30052
M53, *Microfiltration and Ultrafiltration Membranes for Drinking Water*, #30053
M54, *Developing Rates for Small Systems*, #30054
M55, *PE Pipe—Design and Installation*, #30055
M56, *Nitrification Prevention and Control in Drinking Water*, #30056
M57, *Algae: Source to Treatment*, #30057
M58, *Internal Corrosion Control in Water Distribution Systems*, #30058
M60, *Drought Preparedness and Response*, #30060
M61, *Desalination of Seawater*, #30061
M62, *Membrane Applications for Water Reuse*, #30062
M63, *Aquifer Storage and Recovery*, #30063

M65, *On-Site Generation of Hypochlorite,* #30065

M66, *Cylinder and Vane Actuators and Controls—Design and Installation,* #30066

M68, *Water Quality in Distribution Systems,* #30068

M69, *Inland Desalination and Concentrate Management,* #30069

M77, *Condition Assessment of Water Mains,* #30077